ID0984924

Digital Signal Processing

J. Benesty, T. Gänsler, D. R. Morgan, M. M. Sondhi, S. L. Gay
Advances in Network and Acoustic Echo Cancellation

Springer

Berlin
Heidelberg
New York
Barcelona
Hong Kong
London
Milan
Paris
Singapore
Tokyo

J. Benesty · T. Gänsler · D. R. Morgan
M. M. Sondhi · S. L. Gay

Advances in Network
and Acoustic Echo Cancellation

With 70 Figures

Springer

Series Editors

Prof. Dr.-Ing. Arild Lacroix
Johann-Wolfgang-Goethe-Universität
Institut für angewandte Physik
Robert-Mayer-Str. 2-4
D-60325 Frankfurt

Prof. Dr.-Ing. Anastasios Venetsanopoulos
University of Toronto
Dept. of Electrical and Computer Engineering
10 King's College Road
M5S 3G4 Toronto, Ontario
Canada

Authors

Jacob Benesty
e-mail: jbenesty@bell-labs.com

Tomas Gänsler
e-mail: gaensler@bell-labs.com

Dennis R. Morgan
drrm@bell-labs.com

M. Mohan Sondhi
e-mail: mms@bell-labs.com

Steven L. Gay
slg@bell-labs.com

Bell Labs
Lucent Technologies
700 Mountain Avenue, 2D-518
Murray Hill, NJ 07974
USA

ISBN 3-540-41721-4 Springer-Verlag Berlin Heidelberg New York

Library of Congress Cataloging-in-Publication-Data applied for
Die Deutsche Bibliothek - CIP-Einheitsaufnahme
Advances in network and acoustic echo cancellation / J. Benesty ... -Berlin ; Heidelberg ; New York ;
Barcelona ; Hong Kong ; London ; Milan ; Paris ; Singapore ; Tokyo : Springer, 2001
(Digital signal processing)
ISBN 3-540-41721-4

Springer-Verlag Berlin Heidelberg New York
a member of BertelsmannSpringer Science+Business Media GmbH

http://www.springer.de

© Springer-Verlag Berlin Heidelberg 2001
Printed in Germany

Typesetting: Camera-ready copy from authors
Cover design: de'blik, Berlin
Printed on acid free paper   SPIN: 10830716   62/3020/M – 5 4 3 2 1 0

Preface

For many decades, hybrid devices have been used to connect 2-wire local circuits and 4-wire long distance circuits in telephone lines. This leads to a well known problem, whereby echoes are generated. The delay introduced by telecommunication satellites exacerbated this problem and the need for new methods of echo control soon became obvious. The best solution to date for solving this problem was invented in the 1960s at Bell Labs by Kelly, Logan, and Sondhi, and consists of identifying the echo path generated by the hybrid by means of an adaptive filter, a technique that became known as an *echo canceler*. The echo canceler allowed full-duplex communication which was not possible with older echo suppression techniques.

Later, with the development of hands-free teleconferencing systems, another echo problem appeared; but this time the echo was due to the coupling between the loudspeaker and microphone. It is not surprising that the same solution was proposed to solve this problem, and most of today's teleconferencing systems have an acoustic echo canceler. More recently, attention has been given to the very interesting problem of multichannel acoustic echo cancellation, which leads to more exciting applications that take advantage of our binaural auditory system.

Advances in Network and Acoustic Echo Cancellation is a book on echo cancellation in general and adaptive filtering in particular. Although the topic may sound a little bit old, most of the ideas presented here are new and several of them were never published before. There is no necessary order for reading the chapters of this book. All the chapters were written in such a way that they can be read independently.

Chapter 1 introduces the general subject of echo cancellation, giving a historical perspective and overall background to the problem. This serves to set the stage for the various contemporary topics addressed in the subsequent chapters of this book.

In Chap. 2, we talk about a whole class of normalized least mean square (NLMS) algorithms, called proportionate NLMS, which converge very fast when identifying the sparse impulse responses typically encountered in network echo cancellation.

In Chap. 3, we propose to model the error signal with a robust distribution and deduce from it a robust fast recursive least-squares adaptive algorithm.

We then show how to successfully apply this new algorithm to the problem of network echo cancellation when combined with a double-talk detector.

Chapter 4 presents some ideas on how to efficiently implement network echo cancelers designed to simultaneously handle a large number of channels. We show how the computational requirement can be reduced by a very large factor — perhaps as large as thirty.

Chapter 5 explains why and for what kind of applications multichannel sound is important for telecommunication. We explain the fundamental difference from the single-channel case and study a so-called nonuniqueness problem. We also present conventional multichannel time-domain adaptive algorithms.

Chapter 6 develops a fast normalized cross-correlation (FNCC) method for double-talk detection. This method is independent of echo path gain but relies on some of the computations performed in the fast recursive least-squares algorithm (FRLS) used in the acoustic echo canceler (AEC). We also generalize the algorithms to the multichannel case. The combination of the FNCC detector and a robust FRLS algorithm results in a system with extremely low sensitivity to double-talk, few detection errors, and fast convergence after echo path changes.

Chapter 7 describes a stereo teleconferencing system and the implementation of a stereo echo canceler in subbands. The emphasis here is on some of the more practical issues that must be dealt with in an actual implementation.

In Chap. 8, we introduce a new theory on how to develop a whole class of adaptive filtering algorithms in the frequency domain from a recursive least-squares criterion, with a block size independent of the length of the adaptive filter. Then, we deduce an exact adaptive algorithm in the frequency domain and study the convergence of this generalized algorithm. We suggest a very useful approximation, deduce several well-known algorithms, and give hints on how to choose values for some very important parameters. Finally, we show a rigorous link between the multi-delay filter (MDF) and the affine projection algorithm (APA) and generalize some of these ideas to the multichannel case.

In Chap. 9, we describe a system for echo cancellation and double-talk control based on frequency-domain algorithms. The advantages of this approach are, among others, algorithm stability, fast convergence and tracking, low computational complexity, and simplicity of implementation. In detail, we derive a new way of calculating a statistic for double-talk detection, based on a robust adaptive multichannel frequency-domain algorithm.

In Chap. 10, we show that many well-known variables or equations such as the Kalman gain, the Wiener-Hopf equation, the input signal covariance matrix, and the Schur complements can be explicitly formulated in terms of linear interpolation. Also, the so-called principle of orthogonality is generalized. From this theory, we then give a generalized least mean square algorithm and a generalized affine projection algorithm.

We hope this book will serve as a guide for researchers and developers, as well as students who desire to have a new and fresh look at echo cancellation, including some new ideas on adaptive filtering in general. We also hope it will inspire many of the readers and will be the source of new ideas to come. As Prof. Hänsler once wrote about echo cancellation [63]: "From algorithms to systems—It's a rocky road." So let's make it a safe highway!

Acknowledgments

We would like to thank Eric Diethorn, Gary Elko, and Peter Kroon for carefully reading a draft of this book and offering many useful comments for its improvement.

Bell Laboratories, Murray Hill, January 2001

Jacob Benesty

Tomas Gänsler

Dennis R. Morgan

M. Mohan Sondhi

Steven L. Gay

Contents

1. An Introduction to the Problem of Echo in Speech Communication

1.1 Introduction

With rare exceptions, conversations take place in the presence of echoes. We hear echoes of our speech waves as they are reflected from the floor, walls, and other neighboring objects. If a reflected wave arrives a very short time after the direct sound, it is perceived not as an echo but as a spectral distortion, or reverberation. Most people prefer some amount of reverberation to a completely anechoic environment, and the desirable amount of reverberation depends on the application. (For example, much more reverberation is desirable in a concert hall than in an office.) The situation is very different, however, when the leading edge of the reflected wave arrives a few tens of milliseconds after the direct sound. In such a case, it is heard as a distinct echo. Such echoes are invariably annoying, and under extreme conditions can completely disrupt a conversation. It is such distinct echoes that we will be concerned with in this book.

Echoes have long been the concern of architects and designers of concert halls. However, since the advent of telephony, they have also been the concern of communications engineers, because echoes can be generated *electrically*, due to impedance mismatches at points along the transmission medium. Such echoes are called *line* or *network echoes*.

If the telephone connection is between two handsets, the only type of echoes encountered are line/network echoes. These echoes are not a problem in local telephone calls over twisted-pair copper lines because the echo levels are small, and the echoes, if any, occur after very short delays. However, in a long-distance connection in which the end-to-end delay is nonnegligible, distinct echoes are heard. A significant source of line/network echoes in such circuits is a device called a hybrid, which we discuss briefly in Sect. 1.2.

Echoes at hybrids have been a potential source of degradation in the telephone network for many decades, and many solutions have been devised to overcome them. Of particular interest to us in this book are devices known as adaptive echo cancelers. Interest in such devices arose during the 1960s, in anticipation of long delays introduced by telephone communications via geostationary satellites [118], [80], [117]. As satellite communication gained an ever-increasing share of telephone traffic during the 1970s, considerable development of echo cancelers took place. Their widespread use began around

1980 with the arrival of a very large scale integration (VLSI) implementation [39]. The advent of fiber optics obviated satellite delay. However, more recently, with the growing use of speech coding in the telecommunications network, delay has again become an issue. Another recent development is the delay introduced by packet-based asynchronous transfer mode (ATM), internet protocol (IP), and wireless communications carried over analog lines. All of these modern trends further increase the reliance on echo control.

When the telephone connection is between hands-free telephones or between two conference rooms, a major source of echoes is the acoustic coupling between the loudspeaker and the microphone at each end. Such echoes have been called *acoustic echoes*, and interest in adaptive cancellation of such echoes has attracted much attention during the past two decades. A caveat to the reader is in order at this point. Although we will be dealing with acoustically generated echoes, we will only consider cancellation of these echoes in the *electrical* portion of the circuit. We will not discuss the related, but much more difficult, problem of canceling echoes acoustically, i.e., active noise control [83]. Although single-channel acoustic echo cancelers are already in use, the more difficult problem of multichannel (e.g., stereo) acoustic echo cancellation will doubtlessly arise in future applications involving multiple conference parties and/or superposition of stereo music and other sound effects, e.g., in interactive video gaming. We will discuss recently developed methods for echo cancellation in such applications.

In the next two sections we will briefly discuss the problem of line/network echoes and the adaptive cancellation of such echoes. We refer the reader to review articles [119], [94] for a more detailed account. Besides introducing the reader to the echo problem, this preliminary discussion will also lay the groundwork for the more modern problem of canceling acoustically generated echoes in both single-channel and multichannel applications, which will be discussed in Sects. 1.4 and 1.5, respectively. Conclusions appear in Sect. 1.6.

1.2 Line/Network Echoes

As mentioned in the preceding section, the main source of line/network echoes is the device known as a hybrid. Figure 1.1 illustrates, in a highly simplified manner, the function and placement of hybrids in a typical long-distance telephone connection.

Every conventional analog telephone in a given geographical area is connected to a central office by a two-wire line, called the customer loop, which serves for communication in either direction. A local call is set up by simply connecting the two customer loops at the central office. When the distance between the two telephones exceeds about 35 miles, amplification becomes necessary. Therefore a separate path is needed for each direction of transmission. The hybrid (or hybrid transformer) connects the four-wire part of the circuit to the two-wire portion at each end. With reference to Fig. 1.1, the

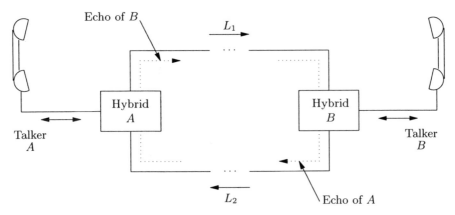

Fig. 1.1. Illustration of a long-distance connection showing local 2-wire loops connected through hybrids to a 4-wire long-line network

purpose of the hybrids is to allow signals from Talker A to go along the path L_1 to B, and to go from Talker B along the path L_2 to A. However, Hybrid B must prevent signals in path L_1 from returning along the path L_2 back to A. Similarly, the signal in path L_2 is to be prevented from returning along path L_1 back to B.

We do not wish to go into the detailed workings of a hybrid. Further information can be found in [119] and other references cited there. Suffice it to say here that a hybrid is a bridge network that can achieve the afore-mentioned objectives, provided the impedance of the customer loop can be exactly balanced by an impedance located at the hybrid. Unfortunately, this is not possible in practice because there are far fewer four-wire circuits than there are two-wire circuits. Therefore, a hybrid may be connected to any of the customer loops served by the central office. By their very nature, customer loops have a wide variety of characteristics — various lengths, type of wire, type of telephone, number of extension phones, etc. It appears, therefore, that the echo at the hybrid cannot be completely eliminated. As a compromise, a nominal impedance is used to balance the bridge, and the average attenuation (in the United States) from input to the return-path output of the hybrid is 11 dB with a standard deviation of 3 dB. This amount of attenuation is not adequate for satisfactory communication on circuits with long delays because the echoes remain audible [119].

1.2.1 The Echo Suppressor

The problem of such echoes has been around ever since the introduction of long-distance communication. On terrestrial circuits, the device most widely used to control line/network echoes is the echo suppressor [119]. Again, we will not describe echo suppressors in detail, but merely mention that they are voice-operated switches whose object is to remove the echo of the talker's

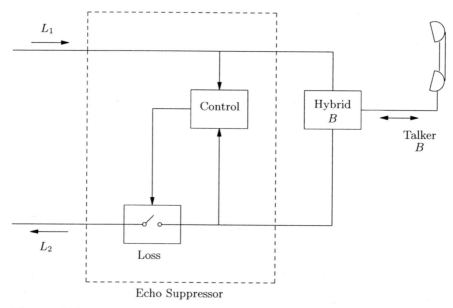

Fig. 1.2. Echo suppressor attempts to remove echo by inserting switched loss when near-end speech (Talker B) is not present

speech and yet allow the listener to interrupt as in normal conversation. The principle of the echo suppressor can be explained by referring to Fig. 1.2, which shows the end B (also known as the "near end") of the telephone circuit of Fig. 1.1, with an echo suppressor included. Suppose the talker at end A (also known as the "far end") has been talking for a while. Based on the level of signals in the paths L_1 and L_2, a decision is made as to whether the signal in L_2 is an interruption by B trying to break into the conversation or an echo of A's speech. If the decision is the latter, the circuit L_2 is opened (or a large loss is switched in). A similar switch at the other end prevents B's echo from returning to B. During so-called "double-talk" periods, when speech from speakers A and B is present simultaneously at the echo suppressor, echo suppression is inhibited so that A hears the speech from B superimposed on self echo from A.

If the decision mechanism were to behave flawlessly, the echo suppressor would be a satisfactory form of echo control. The decision, however, cannot be perfect. The two signals that have to be distinguished are both speech signals, with more or less the same statistical properties. Essentially the only distinguishing property is the level. Therefore, sometimes a high level of echo is returned, and sometimes when the speech level is low (or during initial and final portions of speech bursts) the interrupter's speech is mutilated. However, with considerable ingenuity, echo suppressors have been designed to keep such malfunctions at an acceptable level. Selective echo suppression [95] can also

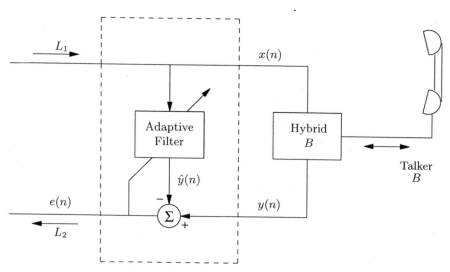

Fig. 1.3. Echo canceler continually removes echo even if near-end talker is active

be applied within the structure of subband echo cancelers, to be discussed later in Sect. 1.4.1.

1.2.2 The Line/Network Echo Canceler

Echo suppressors served well for over 70 years on circuits with round-trip delays of less than about 100 ms, corresponding to land line distances of a few thousand miles. With the advent of commercial communications satellites in 1965, however, the situation changed significantly. A geostationary satellite (i.e., one that is stationary with respect to the earth) must have an orbit that is about 23,000 miles above the earth's surface. A telephone connection via such a satellite will have a round-trip echo delay of 500-600 ms [119]. With such long delays, echo suppressors fail to function satisfactorily. The long delay induces a change in the pattern of conversation in a way so as to significantly increase the number of errors. New methods of echo control were proposed for circuits with such long delays. Of these, the most versatile, and the one in widespread use, is the adaptive echo canceler [118], [80], [117]. The unique feature that makes it so attractive is that, unlike other forms of echo control, the echo canceler does not tamper with the path carrying the echo. Therefore, it never mutilates the speech of the interrupting party. The basic idea of the echo canceler is illustrated in Fig. 1.3. Again we show only one canceler located at the end B of the telephone circuit of Fig. 1.1; a similar canceler is symmetrically located at the other end. As illustrated in Fig. 1.3, instead of interrupting the path L_2, a synthetic echo $\hat{y}(n)$ is generated from A's speech, $x(n)$, and subtracted from the return signal, $y(n)$, to produce the outgoing signal $e(n)$ on the path L_2. Here we assume discrete-time signal processing with time samples denoted by the index n. The synthetic echo is generated by passing $x(n)$ through a filter whose impulse response (or transfer function) matches that of the echo path from $x(n)$ to $y(n)$ via hybrid B.

As mentioned in Sect. 1.2, the echo path is highly variable, so the filter in Fig. 1.3 cannot be a fixed filter. It must be estimated for the particular local loop to which the hybrid gets connected. One simple way to derive the filter is to measure the impulse response of the echo path and then approximate it with some filter structure, e.g., a tapped delay line. However, the echo path is, in general, not stationary. Therefore, such measurements would have to be made repeatedly during a conversation. Clearly this is highly undesirable. To eliminate the need of such measurements the filter is made adaptive. An algorithm is implemented which uses the residual error to adapt the filter to the characteristics of the local loop, and to track slow variations in these characteristics. In the next section we will discuss several basic adaptation algorithms in some detail.

1.3 Adaptive Cancellation

To implement a filter that approximates the echo path, the first step is to choose a representation of the filter in terms of a finite number of parameters. Assuming the echo path to be linear, this can be achieved by finding an expansion of the impulse response of the echo path in terms of a set of basis functions. The problem then reduces to the estimation of the expansion coefficients. If $w_l(n), l = 0, 1, 2, \cdots, L - 1$, is the (truncated) set of basis functions, the expansion can be implemented by the set of L filters illustrated in Fig. 1.4(a). The output of the filter bank, $y_e(n)$, is related to the input $x(n)$ by the relation

$$y_e(n) = x(n) * \sum_{l=0}^{L-1} h_l w_l(n)$$

$$= \sum_{l=0}^{L-1} h_l x_l(n)$$

$$= \mathbf{h}^T \mathbf{x}(n) . \tag{1.1}$$

Here $*$ indicates convolution, $x_l(n)$ is the output of the lth filter, and h_l is the lth expansion coefficient. In the last line of (1.1) we have introduced matrix notation which will be useful later. The boldface quantities $\mathbf{h}$ and $\mathbf{x}$ are column vectors with dimension $L \times 1$, and ^{T} denotes matrix transpose. We note in passing that the gain factors, $h_0, h_1, \ldots, h_{L-1}$ in Fig. 1.4(a) could be generalized to FIR filters; this is the essence of the subband adaptive filter concept to be discussed later in Sect. 1.4.1.

In the special case when $w_l(n) = \delta(n - l)$, the filter becomes an L-tap transversal filter (tapped delay line) with a unit delay between taps, as il-

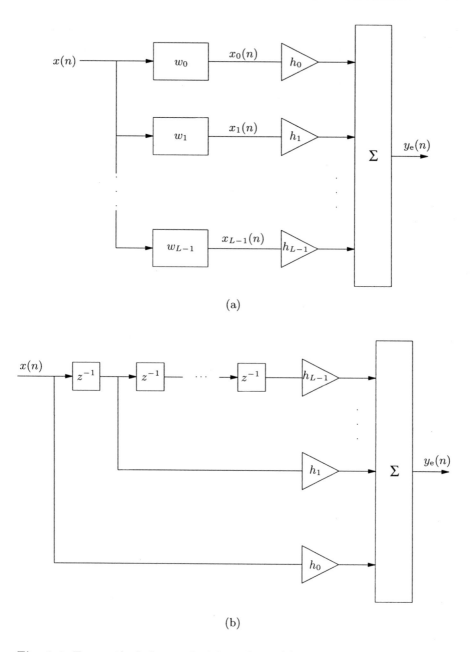

Fig. 1.4. Two methods for synthesizing echoes: (a) using a filter expansion and (b) tapped delay line filter

lustrated in Fig. 1.4(b). This structure is also known as a finite impulse response (FIR) filter. It is the most commonly used filter structure, although other structures, e.g., when the $w_l(n)$'s are Laguerre functions or truncated (or damped) sinusoids, have been tried [117].

The general properties of the adaptation algorithms that we shall discuss presently arise mainly from the fact that the output depends linearly on the parameters h_l. Therefore, our discussion will apply for any choice of functions $w_l(n)$ (although, of course, the rate of convergence will depend strongly on that choice). The general features will, in fact, be valid even if the $x_l(n)$'s are nonlinearly filtered versions of $x(n)$. This fact allows one to handle a class of nonlinear echo paths by the same methods [126].

1.3.1 The Stochastic Gradient Algorithm

By far the most popular algorithm for adapting the filter structure of (1.1) to the echo path is the stochastic gradient algorithm. It is now popularly known as the least mean square (LMS) algorithm and was first introduced around 1960 for adaptive switching [133]. The LMS algorithm was initially used for echo cancelers [117] and adaptive antenna arrays [3] in the mid 1960s. Since then, its use has expanded to the general field of adaptive signal processing [134], [66], finding applications in many other areas, such as interference cancellation, equalization, and system identification.

The basic idea of the stochastic gradient algorithm is quite simple. Suppose $y(n)$ is the hybrid return signal in Fig. 1.3. Let us assume that

$$y(n) = y_e(n) + v(n) + w(n) , \tag{1.2}$$

where $y_e(n)$ is an echo of the input signal $x(n)$, $v(n)$ is possible near-end speech of Talker B, and $w(n)$ an added noise component. We will assume that $y_e(n)$ has the representation given in (1.1) for some (unknown) coefficient vector $\mathbf{h}$. If this is not strictly true, $w(n)$ will include the residual modeling error as well.

Suppose an estimate of the echo

$$\hat{y}(n) = \hat{\mathbf{h}}^T \mathbf{x}(n) \tag{1.3}$$

is formed with a trial coefficient vector $\hat{\mathbf{h}}$. We wish to implement an algorithm to improve $\hat{\mathbf{h}}$, i.e., bring it closer to the vector $\mathbf{h}$. Since $\mathbf{h}$ is unknown, we must evaluate the goodness of $\hat{\mathbf{h}}$ indirectly. One measure of the performance of $\hat{\mathbf{h}}$ is the error

$$e(n) = y(n) - \hat{y}(n) . \tag{1.4}$$

Since the objective is to make the vector $\hat{\mathbf{h}}$ approximate the vector $\mathbf{h}$, one might search for the vector $\hat{\mathbf{h}}$ that minimizes the expected value of the squared error $e^2(n)$. A natural way is to move $\hat{\mathbf{h}}$ in the direction opposite to the

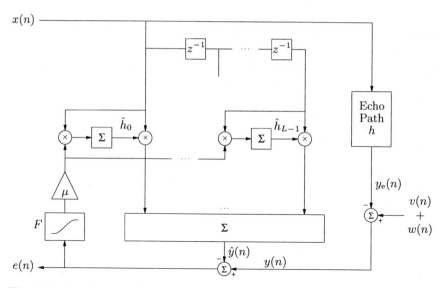

Fig. 1.5. An echo canceler utilizing the stochastic gradient technique, also known as the LMS algorithm

gradient of this expected error. Thus, one might try to increment $\hat{\mathbf{h}}$ by the value

$$\Delta\hat{\mathbf{h}} = -\frac{\mu}{2}\nabla E\{[y(n) - \hat{y}(n)]^2\}$$

$$= -\frac{\mu}{2}\nabla E[e^2(n)] , \tag{1.5}$$

where μ is a parameter that controls the rate of change, E denotes mathematical expectation, and ∇ is the gradient with respect to $\hat{\mathbf{h}}$. Equation (1.5) is just one form of gradient search for the location of the minimum of a function of several variables. What the stochastic gradient algorithm does is to replace the *expected value* of the squared error by the *instantaneous value*. As we shall see, even such a crude estimate of the gradient is adequate, under certain reasonable conditions, to make $\hat{\mathbf{h}}$ approach $\mathbf{h}$.

The stochastic gradient version of (1.5) is

$$\Delta\hat{\mathbf{h}} = -\frac{\mu}{2}\nabla[e^2(n)]$$

$$= -\mu e(n)\nabla[e(n)]$$

$$= \mu e(n)\mathbf{x}(n) . \tag{1.6}$$

Therefore, at time index n, we update the adaptive weight as

$$\hat{\mathbf{h}}(n + 1) = \hat{\mathbf{h}}(n) + \mu\mathbf{x}(n)e(n) . \tag{1.7}$$

Figure 1.5 illustrates the block diagram of an algorithm to implement the adaptation according to (1.7). The diagram includes a function F that equals the identity function when implementing (1.7). The introduction of F allows one to handle a more general criterion than the squared error [117]. For instance, if the expectation of the *magnitude* of the error is to be minimized, $F(\cdot) = \text{sign}(\cdot)$ must be chosen. This choice of F has been used in some implementations of the algorithm, (see, e.g., [122], [127]). Another useful choice is the ideal limiter, which when combined with an adaptive scale factor leads to a class of robust echo cancelers [48], [51]. This topic will be discussed at length in Chaps. 2, 3, 6, and 9.

Convergence in the Ideal Case. Suppose first that the echo path is perfectly stationary, the model represents the echo path exactly, and there is no noise or interrupting speech. Under these ideal conditions we have $y(n) = \mathbf{h}^T\mathbf{x}(n)$, and the error $e(n)$ is given by

$$e(n) = [\mathbf{h} - \hat{\mathbf{h}}(n)]^T\mathbf{x}(n)$$
$$= \boldsymbol{\varepsilon}^T(n)\mathbf{x}(n) , \tag{1.8}$$

where

$$\boldsymbol{\varepsilon}(n) \equiv \mathbf{h} - \hat{\mathbf{h}}(n) \tag{1.9}$$

is the misalignment vector. Rewriting the stochastic gradient algorithm (1.7) in terms of the misalignment gives

$$\boldsymbol{\varepsilon}(n+1) = \boldsymbol{\varepsilon}(n) - \mu\mathbf{x}(n)e(n) . \tag{1.10}$$

Taking the l_2 norm of both sides and using (1.8), we then obtain

$$\|\boldsymbol{\varepsilon}(n+1)\|^2 - \|\boldsymbol{\varepsilon}(n)\|^2 = -\mu[2 - \mu\|\mathbf{x}(n)\|^2]e^2(n) . \tag{1.11}$$

This shows that for $\mu < 2/\|\mathbf{x}(n)\|^2$, the length of the misalignment vector $\boldsymbol{\varepsilon}$ is nonincreasing. It is strictly decreasing as long as there is an uncanceled echo. Another important piece of information that can be gathered from (1.11) is that $e^2(n)$ eventually goes to zero; this is seen by summing both sides of (1.11) from $n = 0$ to $N - 1$, which yields

$$\|\boldsymbol{\varepsilon}(0)\|^2 - \|\boldsymbol{\varepsilon}(N)\|^2 = \mu\sum_{n=0}^{N-1}[2 - \mu\|\mathbf{x}(n)\|^2]e^2(n) . \tag{1.12}$$

Since the left-hand side is bounded by the initial value $\|\boldsymbol{\varepsilon}(0)\|^2$, it follows that for $\mu < 2/\max_n \|\mathbf{x}(n)\|^2$, $e^2(n)$ must, in the limit, go to zero.

We cannot, however, be satisfied with the error going to zero; we want the error to be zero not only for the signal history up to the present, but for *all* subsequent signals. Hence, what we require is that the misalignment vector $\boldsymbol{\varepsilon}(n)$ should go to zero. Unfortunately, that is not provable even in the ideal situation considered in this section without imposing conditions on the input signal $x(n)$. The reason is that $e(n) = 0$ does not imply that $\boldsymbol{\varepsilon}(n) = 0$, but only that $\boldsymbol{\varepsilon}(n)$ is orthogonal to $\mathbf{x}(n)$.

Sufficient conditions for the convergence of $\boldsymbol{\varepsilon}(n)$ to zero, intuitively speaking, assure that the time varying vector $\mathbf{x}(n)$ does not stay confined to a subspace of dimension less than L for too long; i.e., $\mathbf{x}(n)$ should evolve in time in such a way as to cover the entire L-dimensional space. This condition is commonly referred to as "persistent excitation" [66, pp. 748-751].

Even when the persistent excitation condition is satisfied, it is a difficult matter to get accurate estimates of the *rate* of convergence of $\boldsymbol{\varepsilon}(n)$. Suppose, for instance, that $x(n)$ is a member of a stationary ergodic process. One would expect that in this case the expected convergence rate could be easily computed. This is not the case. If, for instance, the expectation of both sides of (1.11) is taken, it does not help because $e(n)$ on the right-hand side depends on $\boldsymbol{\varepsilon}(n)$ itself. However, if μ is very small, one can assume that $\boldsymbol{\varepsilon}(n)$ and $\mathbf{x}(n)$ are independent: $\boldsymbol{\varepsilon}(n)$ changes slowly, and the expectation on the $\mathbf{x}(n)$ ensemble, E_x, can be taken with $\boldsymbol{\varepsilon}(n)$ assumed quasi-constant. This is known as the *independence assumption* [117], which can be justified rigorously as a first-order perturbation approximation [92]. Under this assumption we see that, using (1.8),

$$
\begin{aligned}
E_x[e^2(n)] &= E_x[\boldsymbol{\varepsilon}^T(n)\mathbf{x}(n)\mathbf{x}^T(n)\boldsymbol{\varepsilon}(n)] \\
&= \boldsymbol{\varepsilon}^T(n)\mathbf{R}\boldsymbol{\varepsilon}(n) ,
\end{aligned}
\tag{1.13}
$$

where $\mathbf{R} \equiv E[\mathbf{x}(n)\mathbf{x}^T(n)]$ is the correlation matrix of $\mathbf{x}(n)$. Then taking the expected value of (1.11), assuming that the step size is small enough so that μ^2 can be neglected, and using (1.13) gives the geometrically decaying upper and lower bounds

$$
(1 - 2\mu\lambda_{\max})^n \leq E[\|\boldsymbol{\varepsilon}(n)\|^2] \leq (1 - 2\mu\lambda_{\min})^n ,
\tag{1.14}
$$

where $\lambda_{\max}$ and $\lambda_{\min}$ are, respectively, the maximum and minimum eigenvalues of $\mathbf{R}$.

Fortunately, for convergence rates of interest in general, these bounds are useful. Nevertheless, it is important to remember that the bounds are *not* valid for large μ. For example, if $2\mu\lambda_{\min} = 1$, the bounds are meaningless. Furthermore, if the step size μ is increased beyond a certain value, the error will diverge. Equation (1.14) suggests a stability limit $\mu < 1/\lambda_{\max}$. However, we recall that this result is based on the assumption of a sufficiently small step size μ. Although this condition is obviously not met here, the resulting step-size limit is not too far off. For a Gaussian process, it can be shown [46] that stability is assured for

$$
\mu < \frac{2/3}{\text{tr}(\mathbf{R})} \leq \frac{2/3}{\lambda_{\max}} .
\tag{1.15}
$$

Although the convergence rates are difficult to estimate, it is clear from (1.14) that the convergence rate can fluctuate quite a lot if the spread of eigenvalues of the correlation matrix $\mathbf{R}$ is large. To reduce these fluctuations, one would ideally want to "whiten" the speech signal (i.e., make all the eigenvalues equal and constant). Speech is a nonstationary signal whose spectral properties change much more rapidly than does the echo path. Whitening it, therefore, requires a fast adaptation, in addition to the adaptation to the echo path. However, one source of variability of the eigenvalues can be eliminated rather easily. This is the variability due to the change in signal level. Since the eigenvalues are proportional to the variance (or power) of the input signal, this objective can be accomplished by dividing the right-hand side of (1.6) by a local estimate of power. Thus, (1.7) is modified to

$$\hat{\mathbf{h}}(n+1) = \hat{\mathbf{h}}(n) + \mu \frac{e(n)}{\mathbf{x}^T(n)\mathbf{x}(n)} \mathbf{x}(n) \ , \tag{1.16}$$

where now μ is the normalized step size. Because of the division by the input power, this algorithm is called the *normalized* LMS (NLMS) algorithm.

The NLMS algorithm of (1.16) can be interpreted as a projection that solves an underdetermined least mean square problem [66, pp. 352-356], and for a stationary process, convergence in the first and second moment is guaranteed for $\mu < 2$ [115]. This result has also been shown to hold for a (nonstationary) spherically invariant process [110], which has been suggested as a model for speech signals. For both LMS and NLMS, a good rule of thumb for achieving fastest convergence is to set the step size to about half of its maximum stable value, i.e., $\mu \approx (1/3)/\mathrm{tr}(\mathbf{R})$ for LMS and $\mu \approx 1$ for NLMS. However, in practice, smaller values are usually used to insure stability in the presence of transient disturbances.

Convergence in the Nonideal Case. The convergence process, in practice, is even more complicated than that described in the previous section. Detailed discussion of the nonideal case is beyond the scope of this book. If the only perturbation is an additive noise, the vector $\boldsymbol{\varepsilon}(n)$ converges to lie within a sphere around the origin, whose radius is proportional to the root-mean-square (rms) value of the noise [132]. If the echo path is not constant, the radius of the sphere is also proportional to the rate of change of the impulse response.

The most severe situation arises during intervals of double talking (i.e., intervals during which the speech from near-end and far-end talkers is present simultaneously at the echo canceler). If the echo canceler has converged to a small misalignment, the interfering near-end speech signal, $v(n)$ in Fig. 1.5, can be much louder than the uncancelled echo, $e(n)$, and can completely misalign the canceler in a very short time. About the only effective way of dealing with this problem is to use a system similar to the echo suppressor (see Sect. 1.2.1) to detect the occurrence of double talking. However, instead of breaking the return path, just the adaptation loop is temporarily disabled during these intervals.

One of the most widely used double-talk detectors is the so-called "Geigel algorithm" [37], which declares the presence of near-end speech whenever

$$|y(n)| > T \max_{n-L \leq m < n} |x(m)| \, , \tag{1.17}$$

where T is a suitably chosen constant, e.g., $1/2$ (–6 dB). The Geigel algorithm works fairly reliably for line/network echo cancelers where the hybrid echo return loss is reasonably well defined. However, for acoustic echo cancelers, selection of the threshold value T is more problematic because, for example, the return loss may be negative (a gain). Some recent advances in double-talk detection additionally make use of the correlation [140] or coherence [52] between $x(n)$ and $y(n)$ or between $x(n)$ and $e(n)$. A proof of the equivalence of these techniques as well as a new normalized cross-correlation technique appears in [16]. This book will develop forms of the normalized cross-correlation DTD in both the time domain (Chap. 6) and frequency domain (Chap. 9).

1.3.2 Other Algorithms

Before turning to a discussion of acoustic echo cancelers, let us briefly discuss five algorithms that attempt to improve on the simple stochastic gradient algorithm discussed so far: (1) the proportionate NLMS (PNLMS) algorithm, wherein the step size is proportional to the tap weight, (2) a canceler based on two echo-path models, (3) the least-squares (LS) algorithm, (4) the recursive least-squares (RLS) algorithm, and (5) the affine projection algorithm (APA). These are discussed in detail below. However, at this point we can make some cursory comments. The PNLMS algorithm has recently been introduced in the telephone network. However, none of the others is yet in common use for line (or acoustic) echo cancellation, although prototypes have been implemented. The main reason why the two-path approach has not found wide application is the difficulty in designing a decision algorithm needed for its implementation, as well as the additional memory requirements. As for the LS and RLS algorithms, the main reason is the added computational complexity. The APA attempts to bridge the gap in complexity between the LS and RLS algorithms and the much simpler LMS and NLMS algorithms. Also, there are now fast recursive least-squares (FRLS) and fast affine projection (FAP) algorithms. In view of the rapid advancement of digital technology, these and other more complex algorithms will, no doubt, be used in the near future. As a final note, we would add that in face of the ever increasing network delay due to modern communication system trends, efficient computational techniques used in acoustic echo cancellation become more and more applicable to line/network echo cancelers. These techniques include subband adaptive filters and frequency-domain algorithms, which will be discussed later in Sect. 1.4.

Proportionate NLMS Algorithm. The proportionate NLMS (PNLMS) algorithm [38] has been developed for applications where the class of possible impulse responses covers a wide time interval, but any particular realization is significant only over a small (but unknown) subset of the taps. This class of responses is said to be *sparse*, and the PNLMS algorithm exploits this property to accelerate the convergence rate. If we knew somehow which small set of coefficients were significant over the length of the adaptive filter, we would only need to update those coefficients because the others would be small enough to neglect. This is a great advantage because the maximum step size of the NLMS algorithm is limited by the power of the reference signal vector, which scales directly by the number of coefficients (see Sect. 1.3.1). Therefore, fewer active coefficient updates means the overall step size can be increased and therefore faster convergence will be realized. However, of course, we do not know where the support of the sparse impulse response is concentrated, so we have to adaptively derive the step sizes to realize the potential benefit. The PNLMS algorithm does this by adaptively increasing the relative step size of each coefficient *proportionate* to its magnitude. Initially, all coefficients are zero, so some minimum increments must also be specified. These details are all worked out in [38] and are recapitulated in Chap. 2. Chapter 2 then goes on to extend the PNLMS algorithm to a robust version that is more tolerant of undetected double-talk.

Two Echo-Path Models. In the discussion of the stochastic gradient algorithm, we defined the error signal $e(n) = y(n) - \hat{\mathbf{h}}^T(n)\mathbf{x}(n)$, which was used to control the adaptation of the canceling filter $\hat{\mathbf{h}}(n)$. In the usual implementation of the echo canceler, this same error signal is the one sent to the remote station as the echo-free signal. However, this is not imperative. An alternative is to employ a "background" adaptive filter $\hat{\mathbf{h}}_b(n)$ to derive a local error signal $e_b(n)$. The returned signal is still derived as $e(n) = y(n) - \hat{\mathbf{h}}^T(n)\mathbf{x}(n)$, but now the filter $\hat{\mathbf{h}}(n)$ is derived from, but not identical to, $\hat{\mathbf{h}}_b(n)$. In [104], it is argued that this technique can be used to advantage. Since the gradient used in the LMS algorithm is only a very crude estimate of the gradient of the mean-squared error, every adaptive step does not necessarily improve $\hat{\mathbf{h}}_b(n)$. The suggestion in [104] is to monitor both $e_b(n)$ and $e(n)$, and to copy the coefficients of $\hat{\mathbf{h}}_b(n)$ into $\hat{\mathbf{h}}(n)$ whenever $\hat{\mathbf{h}}_b(n)$ performs consistently better than $\hat{\mathbf{h}}(n)$ over some specified time interval. The decision when to transfer coefficients may be based on observation of $e_b(n)$, $e(n)$, $\hat{\mathbf{h}}_b(n)$, $\hat{\mathbf{h}}(n)$, $\mathbf{x}(n)$, and $y(n)$. The added memory and computational requirements have discouraged use of this algorithm in the past, although it has been recently incorporated into a few products. Continued reduction in the cost of digital signal processing and memory should make this algorithm find widespread application. Further discussion and extension of the two echo-path approach is covered in Chap. 7. In Chap. 6, a link is made between the two-path model and a fast version of the normalized cross-correlation DTD.

The Least-Squares Algorithm. In Sect. 1.3.1 we noted that the impulse response $\hat{\mathbf{h}}$ can be estimated by minimizing the expectation of the squared error, $e^2(n)$. The stochastic gradient algorithm sidesteps the problem of estimating the *expected* value of $e^2(n)$ by taking an incremental step in the direction that reduces its *instantaneous* value. Instead of this, the LS algorithm minimizes a better deterministic approximation to the expected value. Specifically, the LS algorithm computes the vector $\hat{\mathbf{h}}$ that minimizes the arithmetic mean of $e^2(m)$ over some range of sampling instants m. Thus, at time index n, we minimize, over a block of M samples, the objective function

$$J(\hat{\mathbf{h}}) = \sum_{m=n-M+1}^{n} e^2(m) . \tag{1.18}$$

Substituting for $e(n)$ in terms of $\hat{\mathbf{h}}(n)$ and $\mathbf{x}(n)$ reduces the problem to minimizing

$$J(\hat{\mathbf{h}}) = \hat{\mathbf{h}}^T(n)\mathbf{X}^T(n)\mathbf{X}(n)\hat{\mathbf{h}}(n) - 2\mathbf{r}^T(n)\hat{\mathbf{h}}(n) + \sum_{m=n-M+1}^{n} y^2(m) , \tag{1.19}$$

where $\mathbf{X}(n)$ is an $M \times L$ matrix whose mth row ($1 \le m \le M$) is $\mathbf{x}^T(n - M + m)$, and the vector

$$\mathbf{r}(n) \equiv \sum_{m=n-M+1}^{n} y(m)\mathbf{x}(m) . \tag{1.20}$$

Setting the gradient of $J(\hat{\mathbf{h}})$ in (1.19) to zero shows that $\hat{\mathbf{h}}(n)$ is the solution of

$$\mathbf{X}^T(n)\mathbf{X}(n)\hat{\mathbf{h}}(n) = \mathbf{r}(n) , \tag{1.21}$$

and can be computed by a single matrix inversion. Since the matrix to be inverted is of size $L \times L$, its inversion would ordinarily require $O(L^3)$ computations. [Recall that L is the length of the vector $\hat{\mathbf{h}}(n)$.] However, if the adaptive structure is a transversal filter, the matrix $\mathbf{X}(n)$ is Toeplitz (i.e., all entries on any diagonal are identical). Taking advantage of this property, the inversion can be performed in $O(L^2)$ operations [89].

The solution depends on both n and M. The matrix $\mathbf{X}^T(n)\mathbf{X}(n)$ is invertible if and only if $\mathbf{X}(n)$ has independent columns. If this is not the case, the solution is not unique, and a pseudoinverse of $\mathbf{X}^T(n)\mathbf{X}(n)$ can be used to select the minimum-norm solution. If the input and the echo path were stationary, the dependence on n would be eliminated, and a single matrix inversion would be required. However, because of the time variation of $\mathbf{h}(n)$, the solution $\hat{\mathbf{h}}(n)$ must be updated as often as the available hardware allows.

The Recursive Least-Squares Algorithm. The least-squares algorithm is a *block* processing algorithm. An optimum estimate of $\mathbf{h}$ is derived from a block of data (of length M in the preceding description). This optimum estimate is assumed to be valid until the next block of data is processed to give a new estimate of $\mathbf{h}$, and so on. There is an alternative algorithm in which an optimal estimate of **h** is obtained recursively at *every* time instant. The algorithm is a deterministic version of the Kalman filter. At every instant, the estimate $\hat{\mathbf{h}}(n)$ minimizes a weighted sum of the squared errors at all past instants of time. In order to be able to track slowly varying impulse responses, the weighting is chosen such that errors in the remote past do not affect the current estimate. Recursive algorithms can be derived for several weighting functions that achieve this objective. One convenient error measure is

$$J(\hat{\mathbf{h}}) = \sum_{m=-\infty}^{n} \lambda^{n-m} e^2(m) , \qquad (1.22)$$

with λ chosen in the range $0 < \lambda < 1$. The value chosen for λ determines the effective duration of the past input that is used to derive the current estimate $\hat{\mathbf{h}}(n)$.

In terms of $\mathbf{x}(n)$ and $\hat{\mathbf{h}}(n)$, (1.22) can be re-written as

$$J(\hat{\mathbf{h}}) = \hat{\mathbf{h}}^T(n)\mathbf{R}(n)\hat{\mathbf{h}}(n) - 2\mathbf{r}^T(n)\hat{\mathbf{h}}(n) + \sum_{m=-\infty}^{n} \lambda^{n-m} y^2(m) , \qquad (1.23)$$

where the matrix

$$\mathbf{R}(n) \equiv \sum_{m=-\infty}^{n} \lambda^{n-m} \mathbf{x}(m)\mathbf{x}^T(m) \qquad (1.24)$$

and the vector

$$\mathbf{r}(n) \equiv \sum_{m=-\infty}^{n} \lambda^{n-m} y(m)\mathbf{x}(m) . \qquad (1.25)$$

Thus at time n, the optimal impulse response vector $\hat{\mathbf{h}}(n)$ is the solution of

$$\mathbf{R}(n)\hat{\mathbf{h}}(n) = \mathbf{r}(n) . \qquad (1.26)$$

From the definitions of $\mathbf{R}(n)$ and $\mathbf{r}(n)$ it is straightforward to show that they satisfy the recursions

$$\mathbf{R}(n) = \lambda\mathbf{R}(n-1) + \mathbf{x}(n)\mathbf{x}^T(n) \qquad (1.27)$$

and

$$\mathbf{r}(n) = \lambda\mathbf{r}(n-1) + y(n)\mathbf{x}(n) . \qquad (1.28)$$

Because of the recursion in (1.27), $\mathbf{R}^{-1}(n)$ can be obtained by updating $\mathbf{R}^{-1}(n-1)$ through use of the matrix inversion lemma [66, p. 480]. The optimal estimate $\hat{\mathbf{h}}(n)$ is thus obtained recursively from $\hat{\mathbf{h}}(n-1)$. The update algorithm is rather cumbersome, although simple in principle. We refer the reader to [66, ch. 13] for details.

Recursion based on (1.27) and (1.28) requires $O(L^2)$ operations per iteration. Although much less than the $O(L^3)$ computations that would be required without the recursions, the computational load is still much more than the $2L$ multiplications per iteration required by the LMS algorithm. The advantage gained by the extra computations is the highly improved convergence rate. We note, however, that the tracking performance of the RLS algorithm may not be improved over that of the LMS algorithm [42].

As in the case of the LS algorithm, the computational requirements can be reduced dramatically if the adaptive structure is a transversal filter. Algorithms that accomplish this, known as fast RLS (FRLS) or fast transversal filter algorithms, have been developed in [85], [43], [29], and others. They achieve the good convergence properties of the RLS algorithm with a computational requirement that grows linearly with L.

The main limitation of these fast algorithms is that they tend to be numerically unstable unless multiple precision arithmetic is used. In practical implementations the algorithms have to be periodically reset. Nevertheless, as discussed later in Sect. 1.4.1, the fast transversal filter algorithm has been successfully implemented for subband acoustic cancellation. Further progress on stabilization of the fast RLS algorithm has been reported in [116]. Chapter 3 will discuss fast RLS algorithms that are robust to double-talk, and Chap. 5 introduces multichannel RLS algorithms.

Affine Projection Algorithms. An algorithm was introduced in [106] based on affine projections of the most recent M data vectors, and is the basis for algorithms that converge rapidly for autoregressive (AR) processes of order less than or equal to M and with numerical complexity ML, which is intermediate between that of LMS and RLS. A regularized block update version was independently proposed in [82].

The affine projection algorithm (APA) can be viewed as a generalization of the NLMS algorithm ($M = 1$), and can be embellished and interpreted from various viewpoints [100]. A numerically efficient implementation of the algorithm appears in [56], [125]. A frequency-domain APA is presented in Chap. 8 and the multichannel APA is discussed in Chap. 5.

1.4 Single-Channel Acoustic Echo Cancellation

The problem of canceling acoustic echoes in hands-free telephony and teleconferencing differs from the cancellation of line/network echoes mainly because of the different nature of the echo paths [62]. Instead of the mismatch of the hybrid, a loudspeaker-room-microphone system needs to be modeled in these applications (Fig. 1.6). As with line/network echoes, echo suppressors can be

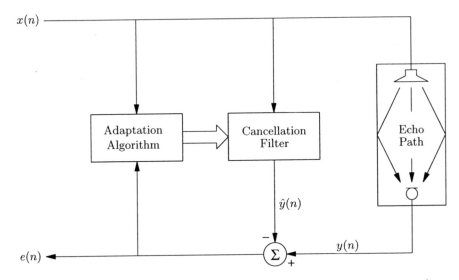

Fig. 1.6. An acoustic echo canceler is used to cancel echoes that arise from coupling between a loudspeaker and microphone

employed (Sect. 1.2.1), but for reasons discussed below are even less satisfactory in this regime. Accordingly, this section will focus mainly on acoustic echo cancellation.

Comparing typical impulse responses of echo paths for line/network echoes and acoustic echoes, it becomes obvious that acoustic echo cancellation is a far more challenging task than line/network echo cancellation. The effective duration of the impulse response of the acoustic echo path is usually several times longer (typically 50-300 ms) and it may change rapidly at any time (e.g., due to an opening door or a moving person). To achieve even a modest improvement, say a misalignment 20 dB below the uncanceled room response, typically requires a transversal filter with 500 taps at an 8-kHz sampling rate for a small office. Higher cancellation down to background levels can easily require thousands of taps at an 8-kHz sampling rate [62]. Some commercial products foresee requiring as many as 4000 taps at a 16 kHz sampling rate.

For line/network echo cancellation, it is sometimes possible to switch to a slower-converging tracking mode after an initial rapid convergence (with, for example, a training signal). This is because for certain circuits it is possible to assume that the echo path varies only very slowly. As previously mentioned, this is not a good assumption for acoustic echo cancellation, so one cannot switch to a slowly converging algorithm even in the tracking mode. Indeed, some theoretical models cast doubt as to whether the LMS algorithm can ever satisfactorily track typical acoustic changes in a room [81], and as previously mentioned, the RLS algorithm may fare even worse in this respect [42].

We see, therefore, that acoustic echo cancelers require more computing power than line echo cancelers for two reasons: first, just because of the length of the impulse response to be modeled, and second, because faster converging/tracking algorithms are desirable. Current products (and prototypes) only partly meet these requirements. This is mainly because they use the LMS type of adaptation algorithms. Such algorithms are known to perform poorly for long impulse responses and with speech as the input signal [66].

As mentioned in Sect. 1.3.2, LS and RLS algorithms can increase convergence speed. They have recently been successfully used to implement line/network echo cancelers. However, they are still infeasible for acoustic echo cancellation because of the long impulse responses involved. Other methods have therefore been explored to improve the convergence speed of LMS type algorithms. One direction aims at "whitening" the speech signal for the adaptation. A simple way to do this is to employ a continuously updated first-order linear predictor to prewhiten the reference signal x used to update the adaptive weight vector [138]. This leads to faster convergence with little increase in complexity. Another direction foresees time-varying step size factors for the different taps of the LMS-adapted FIR filter. A general method for doing this, not specifically tailored for the acoustic echo cancellation problem, was suggested in [64]. Another approach exploits the structure of the impulse response of the acoustic echo path and assigns different step sizes to different sections of the echo path impulse response [117], [87]. The idea is that, ideally, the impulse response samples with large values are adapted with large step size while those with small values get a small step size. (This concept is similar to the principle of the PNLMS algorithm discussed in Sect. 1.3.2.) This should result in a faster overall convergence. Obviously, the efficiency of this method depends highly on the *a priori* knowledge concerning the current echo path impulse response and the ability to adjust the step size accordingly.

As previously mentioned, echo suppressors by themselves are unsatisfactory for acoustic echoes. However, such techniques can be used to suppress the residual echo of an acoustic echo canceler and thus improve performance. These techniques include the concept of "center clipping" [95] and frequency-selective suppression using low-order adaptive transversal filters [91], [128]. Selective echo suppression [95] can also be applied within the structure of subband acoustic echo cancelers, to be discussed later.

All of the aforementioned approaches attempt to improve convergence speed without adding too much computational complexity. Recently, subband techniques have been developed to reduce the computational complexity of acoustic echo cancelers with long impulse responses, while at the same time providing more favorable circumstances for fast convergence. Frequency-domain algorithms can also be used to increase convergence rate and reduce computational complexity. These two approaches are now discussed more fully below.

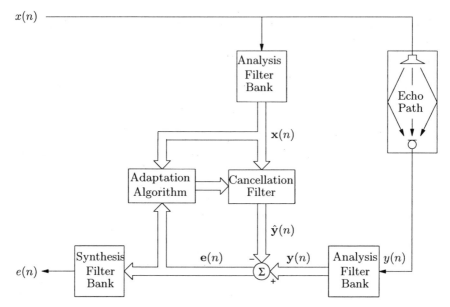

Fig. 1.7. Block diagram of a subband echo canceler showing analysis and synthesis filter banks, subband cancellation filter, and adaptation algorithm

1.4.1 Subband Approach

Subband structures were first proposed in 1984 independently in [47] for teleconferencing purposes, and in [76] as a general framework for acoustic echo cancellation. The fundamental structure is obtained by using bandpass filters for the w_0, w_1, ... , w_{L-1} basis functions in Fig. 1.4(a) and replacing the fixed gains h_0, h_1, ... , h_{L-1} by FIR filters. This leads to the structure depicted in Fig. 1.7, where the wide lines denote multiple subband signals. With reference to that figure, it is seen that the input signal $x(n)$ and echo path output signal $y(n)$ are passed through identical analysis filter banks, producing vectors of M subband signals that are sampled at a reduced rate. The cancellation filter forms a vector of subband signals $\hat{\mathbf{y}}(n)$ to approximate the corresponding subband echo signals $\mathbf{y}(n)$. The resulting subband errors $\mathbf{e}(n)$ are passed through a synthesis filter bank to give a full-band signal $e(n)$, which is then transmitted back to the remote loudspeaker. The adaptation algorithm utilizes the vector of subband error signals and input signals to adjust the input-output characteristics of the cancellation filter bank so as to drive the vector of error signals toward zero.

Note that in Fig. 1.7, each subband echo component is individually cancelled before the composite error signal is synthesized. Alternatively, it is possible to synthesize the estimated full-band echo signal $\hat{y}$ *before* cancellation and use the full-band error signal $e(n)$ to adapt the weights in each subband [78]. However, the synthesis filter delay has the effect of slowing convergence. More importantly, the components of the error signal outside the frequency range of each filter act as a noise on the adaptation process for that filter [120, p. 347]. This requires the step size to be reduced, leading to even slower convergence. Accordingly, we do not consider such structures here.

Intuitively, the promise of the subband structure is twofold: the computational complexity is reduced because of downsampling, and the convergence is accelerated for LMS-type algorithms because the subband signal spectra are more nearly uniform than the full-band signal spectrum. To illustrate the advantage in computational complexity, consider the case in which the canceler has M parallel adaptive filters, and all subband signals are downsampled by the same factor R. Assume that the impulse response of each of the subbands has the same duration as the full-band impulse response. Then, the number of coefficients in each band needed for its representation is fewer by a factor R compared to the full-band representation because of downsampling. Further, filtering and adaptation of the subband cancelers is performed at the reduced sampling rate. Therefore, the computational complexity (measured as computations per second) for one subband canceler is $1/R^2$ that of the full-band canceler. Taking into account all M subbands, the complexity can be expected to be reduced by a factor of approximately R^2/M, assuming that the computational load for the analysis and synthesis systems is negligible. Thus, as R approaches the Nyquist limit M (critical sampling), the computational reduction also approaches a limiting value of M. The reduction in computational complexity can be exploited in several ways: the overall system bandwidth or the duration of the impulse response to be modeled can be increased, more complex adaptation algorithms can be employed, or, most obviously, hardware can be saved to reduce cost. In addition, the subband structure naturally lends itself to efficient parallel processing. The price paid for these advantages can also be inferred from a comparison of Figs. 1.6 and 1.7: in the subband canceler the microphone signal is delayed before it is sent to the far end by the group delay of the cascade of analysis and synthesis filters. Means for dealing with this problem of delay will be described in the last part of this section.

For analysis and synthesis we consider the general class of systems that can be represented as filter banks with time-invariant filtering and downsampling or upsampling, respectively (see, e.g., [130], [129]). Much work has been carried out to develop cancelers using only two or four bands [47], [139]. However, other authors have tried to explore the concept to a greater extent [78], [27], [79], [65], [59], and [44]. The previously mentioned whitening technique can also be applied in subbands [139].

It is shown in [78] that subband solutions exist if and only if the analysis produces aliasing-free subband signals. Obviously, this requires bandpass filters with infinite stopband attenuation. Also, the downsampling factor R must be chosen such that the passband and transition region of the modulated versions of the analysis bandpass do not overlap in the frequency domain after downsampling. If this requirement is met, each subband canceler has to model an ideally bandpass-filtered and downsampled version of the full-band echo path impulse response. The passband of the ideal bandpass should cover the transition and passband region of the corresponding analysis bandpass. However, this solution can obviously not be realized in the strict sense. First, bandpass filters with infinite stopband attenuation can only be approximated and, second, filtering a finite impulse response with an ideal bandpass filter leads to an impulse response that extends from $-\infty$ to $+\infty$ in the time domain and, therefore, is not realizable either. Thus, two approximation problems are linked to the realization of the frequency-subband concept, and it is important to consider the extent to which the approximation errors impair its usefulness.

Approximation of the Subband Solution. The misalignment of the whole structure has two components: the misalignment due to the residual aliasing within the subband signals and the misalignment of the subband cancelers due to the truncation of their impulse responses. The overall system misalignment is dominated by the truncation effect until it reaches a certain level. For smaller misalignment, the truncation errors of the subband cancelers have little impact and the misalignment of the whole system is determined mainly by the aliasing caused by the finite stopband attenuation of the analysis filter. The reason for this becomes obvious from the following consideration of the spectral distribution of the truncation error.

If only a small number of coefficients is used to approximate the response of the ideal subband canceler, the truncation will cause more or less uniformly distributed deviations from the ideal frequency response of the subband canceler. Beyond a certain number of coefficients, the truncation error in the inner region of the subband will be small and the misalignment of the subband canceler is mainly determined by the amount of error concentrated at the band edges, where the steep slope has to be approximated. This contribution to the misalignment of the subband canceler, however, has little influence on the misalignment of the overall system. This is because the misalignment due to the subband canceler is heavily weighted down at the band edges by the analysis and synthesis bandpass filters. Thus, once a certain level of truncation error has been achieved at the band edges (with even smaller error in the inner region), there is little to be gained by increasing the length of the subband canceler. This observation is important for an efficient design of the canceler: it implies that the delay needed to model "noncausal" coefficients of the subband canceler impulse response can be kept small. Once the truncation error has been reduced to a low value, the stopband attenuation controls the residual misalignment. The stopband attenuation must be chosen sufficiently high so as to keep the misalignment due to aliasing small.

The effect of the aliasing within the subband signals also explains the problems with the otherwise attractive choice of critical sampling $(R = M)$

[59]. Perfect reconstruction filter banks with critical sampling cause the transition regions of the analysis filters to be aliased into the subband signals. This causes severe misalignment of the overall subband structure [59]. For this reason it is necessary to choose $R < M$ to decrease the aliasing of the transition regions, even though this sacrifices some computational efficiency.

Further elaboration and practical details of subband implementations are discussed in Chap. 7.

Adaptation of the Subband Structure. As pointed out earlier, the canceler in each subband is essentially independent of the others. Therefore, the subband structure can utilize any of the adaptation algorithms developed for full-band cancelers. We will discuss a few of them, emphasizing the differences compared to their application in full-band systems.

To date, the NLMS algorithm is the most widespread adaptation algorithm for both full-band and subband systems. For this algorithm, it has been observed that the initial convergence is faster in the subband canceler than in the full-band implementation. As discussed in Sect. 1.3.1, the convergence speed of the LMS algorithm depends directly on the eigenvalue spread of the autocorrelation matrix of the input signal. Therefore, it is often concluded that the eigenvalue spread of the subband signals must be smaller than that of the full-band signal. However, examining the actual eigenvalues shows that their spread is in fact larger for the subband signals than for the full-band signal. This is because the subband signals are not really "whiter." The slopes of the frequency responses of the analysis filters cause notches in the subband spectra at the band edges, thus creating some very small eigenvalues. As a result, the convergence behavior of the subband canceler itself is, in general, not better for a subband signal than for the full-band signal. The improvement that is observed for the subband structure as a whole is the result of the same masking property that affects the truncation errors: the overall system misalignment at the band edges of the subband canceler spectra get little weight due to the characteristics of the analysis and synthesis filters. These parts of the subband canceler spectra are the same ones that are affected by the small eigenvalues of the subband signal, and, therefore, correspond to the slower converging modes. But as these modes have little influence on the overall system misalignment, the subband canceler initially converges faster than a full-band system because the eigenvalue spread in the inner region of the subband spectra is indeed smaller than that of the full-band signal. The increased convergence speed for "nonwhite" input signals is, therefore, the result of an only indirect "whitening" effect. The residual band-edge components, however, cause slow asymptotic convergence [98]. As a solution to this problem, it was suggested in [98] that the analysis filter bandwidth be increased so as to push out the band-edge energy beyond the passband of the synthesis filter, thereby eliminating slowly converging components. This idea was subsequently developed and demonstrated in [32].

Skipping all the incremental improvements that are possible to speed up the normalized LMS algorithm in the subband structure (see Sect. 1.4 on full-band approaches), let us mention briefly the most sophisticated adaptation algorithm used so far for acoustic echo cancellation. This is the subband RLS implementation, as proposed in [65]. This approach is further developed in [40], [41] for the multichannel case, which will be discussed in Chap. 7.

Extensions of the Subband Structure. There are at least two areas in which subband cancelers could be improved. First, the cancelers discussed previously keep aliasing small by choosing a decimation rate R less than the number of filters M. The choice $R = M$ is not recommended because of the spectral gaps needed to avoid aliasing. In an attempt to realize this maximum decimation, a structure is proposed in [59] that uses adaptive cross filters to cancel the influence of aliasing in each subband. However, it appears doubtful that this structure would have satisfactory convergence properties, except, perhaps, in some special cases. The main subband filter and the corresponding cross filters have to use the same subband error signal to steer their respective adaptation by the LMS algorithm. As such, the adaptation algorithms for the different filters have no indication of the contribution of each to the total error. Therefore, it appears to be difficult to achieve stable and fast initial convergence. Also, the fact that additional cross filters have to be adapted reduces the computational efficiency, and the gain over a slightly oversampled system (e.g., $R = 3M/4$) might be marginal.

One undesirable aspect of subband cancelers is the delay introduced into the path from the near-end talker to the remote listener. (The delay in the analysis and synthesis filters shown in the bottom portion of Fig. 1.7.) This delay may be 10-20 ms, giving a round-trip delay which is twice that value. In some applications this much delay may not be tolerable. An interesting variant of the subband canceler, which eliminates this delay, has been proposed in [27], whereby the analysis and synthesis filters are moved from the signal return path to the cancellation filter path, and an extra short full-band canceler is added in parallel to compensate for the unmodeled initial delay. This system is, in principle, able to cancel the echoes without introducing delay. However, there are some disadvantages because of the delay introduced in the error feedback loop and the computational increase due to the auxiliary full-band adaptive filter, which is comparable to that of the subband filters. Another technique to eliminate signal path delay, while retaining the computational advantages of subband processing, was introduced in [101], whereby the adaptive weights are computed in subbands, but are then transformed to an equivalent full-band FIR filter, thereby eliminating any delay in the signal path.

1.4.2 Frequency-Domain Algorithms

Another way of efficiently handling long impulse responses with many taps is to employ frequency-domain techniques. With this method, blocks of input samples can be transformed and processed in the frequency domain using the fast Fourier transform (FFT). This approach has several advantages. For one, faster convergence is generally realized since the step size can be independently optimized for each frequency component. Another advantage accrues for long impulse response models, as, e.g., are encountered in acoustic echo cancellation: multiplication replaces convolution in the adaptive filter, so greater computational efficiency is possible. One disadvantage of this approach is the associated block delay. However, this shortcoming can be moderated by using shorter blocks with multi-tap frequency-domain adaptive filters, a technique known as *multi-delay filtering* (MDF). Chapter 8 presents a general derivation of frequency-domain adaptive filters that includes MDF as a special case. The MDF concept is also closely related to subband processing, as discussed in the last section, and elaborated later in Chap. 7. Other processing functions, like double-talk detection, can also be performed in the frequency domain, as discussed in Chap. 9.

1.5 Multichannel Acoustic Echo Cancellation

Until now, we have only considered single-channel acoustic echo cancellation, which is the most prevalent in current usage. However, there are many new applications in which multichannel sound, e.g., stereo, is envisioned in order to provide an ever more lifelike and transparent audio/video medium. These applications include multiparty room-to-room conferencing, multiparty desktop conferencing, and interactive video gaming involving multichannel sound. In these multichannel applications, there are multiple acoustic paths from multiple loudspeakers to multiple microphones, and echoes arising from these paths must be cancelled for full-duplex communication. As we shall see, there are unexpected complications with multichannel sound that do no not occur in single-channel cancelers.

As an introduction to the fundamental problem of multichannel acoustic echo cancellation, consider the room-to-room stereo conferencing scenario of Fig. 1.8 [121]. A transmission room is depicted on the right, wherein two microphones are employed to pick up signals $x_1(n)$, $x_2(n)$ from a talker via two acoustic paths represented by the impulse responses g_1 and g_2. (For convenience, all acoustic paths are assumed to include loudspeaker and/or microphone responses.) These stereophonic signals are then transmitted to loudspeakers in the receiving room on the left, which in turn are coupled to one of the microphones via the paths indicated with impulse responses h_1 and h_2, producing an outgoing signal $y(n)$. Similar paths couple to the other microphone in the receiving room, but for simplicity, only echo cancellation for the one microphone signal will be discussed here; similar remarks will apply to the other microphone signal.

The previously discussed single-channel acoustic echo canceler is thus generalized using two adaptive FIR filters $\hat{h}_1$ and $\hat{h}_2$ to model the two echo paths

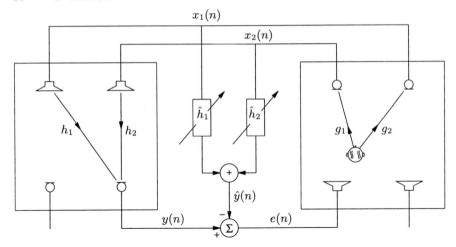

Fig. 1.8. Schematic diagram of stereophonic echo cancellation, showing the use of adaptive filters $\hat{h}_1$ and $\hat{h}_2$ to cancel echo $y(n)$ arising from echo paths h_1 and h_2 from two loudspeakers on the left to one of the microphones

in the receiving room. Driving these filters with the loudspeaker signals $x_1(n)$ and $x_2(n)$ produces an estimate $\hat{y}(n)$ that is subtracted from the echo signal $y(n)$ to form an error signal $e(n)$, which is intended to be small in the absence of near-end speech (i.e., speech generated in the receiving room.)

In order to generalize the LMS algorithm of Sect. 1.3.1 for stereo acoustic echo cancellation, let the echo signal (neglecting near-end speech and ambient room noise) be expressed as [121]

$$y(n) = \mathbf{h}_1^T \mathbf{x}_1(n) + \mathbf{h}_2^T \mathbf{x}_2(n) \,, \tag{1.29}$$

where $\mathbf{h}_1$ and $\mathbf{h}_2$ are L-dimensional vectors of the loudspeaker-to-microphone impulse responses in the receiving room, and

$$\mathbf{x}_1(n) \equiv \begin{bmatrix} x_1(n) \; x_1(n-1) \; \cdots \; x_1(n-L+1) \end{bmatrix}^T$$
$$\mathbf{x}_2(n) \equiv \begin{bmatrix} x_2(n) \; x_2(n-1) \; \cdots \; x_2(n-L+1) \end{bmatrix}^T$$

are vectors comprising the L most recent loudspeaker signal samples. The error signal is then written as

$$e(n) = y(n) - \hat{\mathbf{h}}_1^T(n)\mathbf{x}_1(n) - \hat{\mathbf{h}}_2^T(n)\mathbf{x}_2(n) \,, \tag{1.30}$$

where $\hat{\mathbf{h}}_1$ and $\hat{\mathbf{h}}_2$ are L-dimensional vectors of the adaptive filter coefficients. The error signal can be written more compactly as

$$e(n) = y(n) - \hat{\mathbf{h}}^T(n)\mathbf{x}(n) \tag{1.31}$$

with

$$y(n) = \mathbf{h}^T \mathbf{x}(n) \,, \tag{1.32}$$

where $\hat{\mathbf{h}}(n) \equiv \left[\hat{\mathbf{h}}_1^T(n) \; \hat{\mathbf{h}}_2^T(n) \right]^T$ is the concatenation of $\hat{\mathbf{h}}_1(n)$ and $\hat{\mathbf{h}}_2(n)$, and likewise, $\mathbf{h} \equiv \left[\mathbf{h}_1^T \; \mathbf{h}_2^T \right]^T$ and $\mathbf{x}(n) \equiv \left[\mathbf{x}_1^T(n) \; \mathbf{x}_2^T(n) \right]^T$. In terms of $\mathbf{h}$, we can rewrite (1.31) as

$$e(n) = \left[\mathbf{h} - \hat{\mathbf{h}}(n) \right]^T \mathbf{x}(n) = \boldsymbol{\varepsilon}^T(n)\mathbf{x}(n) , \qquad (1.33)$$

where

$$\boldsymbol{\varepsilon}(n) \equiv \mathbf{h} - \hat{\mathbf{h}}(n) \qquad (1.34)$$

is the composite misalignment vector. With this notation, the two-channel NLMS algorithm can be expressed as

$$\hat{\mathbf{h}}(n+1) = \hat{\mathbf{h}}(n) + \mu \frac{e(n)}{\mathbf{x}^T(n)\mathbf{x}(n)} \mathbf{x}(n) , \qquad (1.35)$$

where μ is the adaptation step size.

For the adjustment of the adaptive filter, (1.35) is formally identical to the adjustment of the single-channel echo canceler discussed in Sect. 1.3.1. It therefore follows that the error $e(n)$ eventually converges to zero for suitable choice of the step size μ. However, $e(n) \to 0$ does not necessarily imply that $\|\boldsymbol{\varepsilon}(n)\| \to 0$ (i.e., $\hat{\mathbf{h}} \to \mathbf{h}$), which is the primary goal of the adaptive filter. The importance of requiring that $\hat{\mathbf{h}} \to \mathbf{h}$ was discussed in Sect. 1.3.1 for the single-channel case; as we will see, this becomes even more critical in the case of multichannel cancelers.

1.5.1 Misalignment: the Nonuniqueness Problem

The main new feature that distinguishes stereo echo cancelers from conventional single-channel cancelers can be explained even without considering the "control" aspects of the adaptation algorithm. Therefore, setting aside the important question of *how* convergence is achieved, let us for the moment just assume that $e(n)$ has been driven to be identically zero. From (1.33), it follows that

$$\varepsilon_1 * x_1 + \varepsilon_2 * x_2 = 0 , \qquad (1.36)$$

where ε_1 and ε_2 are components of the misalignment corresponding to $h_1 - \hat{h}_1$ and $h_2 - \hat{h}_2$, respectively. For the single-talker situation depicted in Fig. 1.8, this further implies

$$[\varepsilon_1 * g_1 + \varepsilon_2 * g_2] * s(n) = 0 , \qquad (1.37)$$

where $s(n)$ is the acoustic signal generated by the talker. In the frequency domain, (1.37) becomes

$$[\mathcal{E}_1(z)G_1(z) + \mathcal{E}_2(z)G_2(z)]S(z) = 0 , \tag{1.38}$$

where the z-transforms of time functions are denoted by corresponding uppercase letters.

Consider first a single-channel situation, say $G_2 = 0$. In that case, except at zeroes of $G_1 S$, (1.38) yields $\mathcal{E}_1 = 0$. Thus, complete alignment ($\hat{h}_1 = h_1$) is achieved by ensuring that $G_1 S$ does not vanish at any frequency.

In the stereophonic situation, on the other hand, even if S has no zeroes in the frequency range of interest, the best that can be achieved is

$$\mathcal{E}_1 G_1 + \mathcal{E}_2 G_2 = 0 . \tag{1.39}$$

This equation does *not* imply $\mathcal{E}_1 = \mathcal{E}_2 = 0$, which is the condition of complete alignment. The problem with stereo echo cancelers is apparent from (1.39): even if the receiving room impulse responses h_1 and h_2 are fixed, any change in G_1 or G_2 requires readjustment of $\mathcal{E}_1$ and $\mathcal{E}_2$, except in the unlikely condition where $\mathcal{E}_1 = \mathcal{E}_2 = 0$. Thus, not only must the adaptation algorithm track variations in the *receiving* room, it must also track variations in the *transmission* room. The latter variations are particularly difficult to track; for if one talker stops talking and another start talking at a different location, the impulse responses g_1 and g_2 change abruptly and by very large amounts. The difficult challenge, then, is to devise an algorithm that (as in the case of a single-channel canceler) converges independently of variations in the transmission room.

We note that the fundamental problem is not resolved even if we know the impulse responses g_1 and g_2 *exactly*, as for example in desktop conferencing where $x_1(n)$ and $x_2(n)$ are synthesized (e.g., synthesized stereo) by appropriate choice of g_1 and g_2. This is because there is still no way to uniquely identify h_1 and h_2.

If two or more independent and spatially separated sources are active in the transmission room, the nonuniqueness problem essentially disappears because (1.39) cannot be simultaneously satisfied for two linearly independent choices of the vector (G_1, G_2) unless $\mathcal{E}_1 = \mathcal{E}_2 = 0$. Similarly, if the transmission room frequency responses G_1 and G_2 vary more rapidly over frequency than $\mathcal{E}_1$ and $\mathcal{E}_2$, this tends to force the misalignment to zero because (1.39) cannot otherwise be simultaneously satisfied over a frequency range for which $\mathcal{E}_1$ and $\mathcal{E}_2$ are constrained to be nearly identical while G_1 and G_2 change appreciably. This situation actually occurs in room-to-room conferencing because the impulse responses g_1 and g_2 are generally longer than L, the number of taps used to model h_1 and h_2, and their frequency responses vary quite rapidly. Thus, the "tails" of the transmission room impulse responses theoretically resolve the nonuniqueness problem; however, solutions so obtained are very poorly conditioned and useless in practice due to the tails of the receiving room impulse responses. This point is more fully addressed in [19] and summarized in Chap. 5.

1.5.2 Search for Solutions

There have been several partially successful attempts to solve the nonunique-ness problem in stereo acoustic echo cancellation. These include the use of a single adaptive filter and various linear signal decorrelation techniques [121], [19]. Single adaptive filters, which attempt to estimate the echo using either $x_1(n)$ or $x_2(n)$ alone, are unsuitable for practical stereo echo cancellation because such a filter still depends strongly on the responses G_1 and G_2 of the transmission room and room responses do not, in general, have stable inverses. Linear signal decorrelation techniques that have not proven satis-factory include: addition of independent random noise to each channel, which is ineffective even if noise shaping is used to exploit masking; use of interchan-nel decorrelation filters, which only resolve ambiguity at frequencies where G_1 or G_2 (but not both) is zero; frequency shifting, which causes the apparent direction of the sound to oscillate, thereby totally destroying the stereophonic effect; and, interleaving comb filters, which have acceptable psychoacoustic degradation only above about 1 kHz.

One solution that has proven effective for speech, is the use of nonlinear distortion in each channel, which has the effect of reducing the coherence between the signals $x_1(n)$ and $x_2(n)$ [19]. This distortion is purposely created by adding to the signal a fraction of its nonlinearly distorted version. Thus, modified signals are formed as

$$x_i'(n) = x_i(n) + \alpha f[x_i(n)] , \ i = 1, 2 , \tag{1.40}$$

where $f(\cdot)$ is a nonlinear function and α is a constant. A choice of f that has proved effective and simple to implement is the half-wave rectifier

$$f(x) = \begin{cases} x , & x \geq 0 \\ 0 , & x < 0 \end{cases} . \tag{1.41}$$

Other choices for the nonlinearity are considered in [99].

The modification (1.40) using the half-wave nonlinearity can also be in-terpreted as the addition of a full-wave nonlinearity, after making suitable scaling changes. The distortion introduced by (1.41), even for values of α as large as 0.3, is hardly noticeable for speech. This is at first surprising because, usually, such high distortion is objectionable in high-fidelity audio systems. One explanation for why speech is not greatly degraded is that the distor-tion for vowel-like sounds is comprised of harmonics that tend to be masked by corresponding harmonics of the original signal. Masking is a well-known psychoacoustic phenomenon by which one sound covers up another, and is also used to advantage in perceptual audio coding.

Even with the use of nonlinear signal transformations, the stereo acoustic echo cancellation problem remains difficult. First of all, two adaptive filters must be used on each end of the communication link, each with two refer-ence signal inputs. Moreover, because the coherence is reduced to only slightly below unity by the nonlinear distortion, the signal components driving convergence to the true solution are relatively small and therefore convergence of the misalignment is very slow with the LMS or NLMS algorithm. For this reason, RLS or other rapidly-converging algorithms (as discussed in Sect. 1.3.2), must be employed for this application. For acoustic echo cancellation involving hundreds or even thousands of taps, such algorithms can consume a great deal of computational power. However, significant computational reduction can be realized by applying FRLS in subbands [40], [41], which will be discussed in Chap. 7.

Chapter 5 discusses other signal decorrelation techniques, develops several fast-converging multichannel adaptive filter algorithms, and presents some practical applications. Chapter 8 includes a frequency-domain approach to multichannel adaptive filtering.

1.6 Concluding Remarks

When adaptive echo cancelers were first proposed over 30 years ago, many people expressed doubts about their economic feasibility and about the feasibility of cancelers with more than 50 or 100 adapted parameters. Advances in digital technology have proven these doubts to be unfounded. Since the appearance of the first VLSI implementation of cancelers in 1980, line echo cancelers have become ubiquitous on the telephone network. Several millions of these devices have now been deployed. Cancelers based on similar principles have also found widespread use in data communication (although we do not consider that application in this book). The most modern application of voice echo cancelers is to the cancellation of acoustic echoes, e.g., for hands-free conference telephony. Hardware implementations based on these proposals have been in use since the mid-1980s. Given the pace of development of digital technology, the next decade may well see widespread use of acoustic echo cancelers for offices and larger conference rooms, for both monophonic and stereophonic sound.

In this chapter, we have introduced the major themes for echo control in speech communication. Along the way, certain topics were keyed to advanced topics covered in the following chapters of this book. These chapters can be read in more or less any order, after the general orientation provided by this introductory chapter.

2. A Family of Robust PNLMS-Like Algorithms for Network Echo Cancellation

2.1 Introduction

In this chapter, we present a family of fast-converging algorithms that are extensions of the proportionate normalized least mean square (PNLMS) algorithm introduced by Duttweiler [38], [54]. This new family of algorithms is based on the affine projection algorithm/normalized least mean square (APA/NLMS) algorithm family [106], [100], [55], [125], [66]. What differentiates the new algorithms from the NLMS and APA algorithms is that they inherit the proportional step-size idea from the PNLMS algorithm, i.e., individually assigned step-sizes to each filter coefficient, where the step sizes are calculated from the previous estimate of the echo path. Because of these individual step sizes, the algorithms achieve a higher convergence rate by using the fact that the active part of a network echo path is usually much smaller (4−8 ms) than the possible echo path range (64−128 ms) that has to be covered. A natural extension of the basic PNLMS algorithm is a proportionate affine projection algorithm (PAPA) [51]. This algorithm (family) combines the fast converging APA with the proportional step size technique of PNLMS.

Although the complexity of PNLMS is not significantly higher than that of NLMS ($\approx$ 2 times), it is of interest to reduce the complexity and numerical dynamic range of the algorithm for practical applications. One algorithm that has numerical and complexity advantages is the so-called signed regressor (SR) PNLMS algorithm which will also be presented in this chapter.

Another important factor in the overall echo canceler performance, is its behavior during double-talk, i.e., when the speech of the talkers at each end of the connection arrives at the canceler simultaneously. This conversational mode perturbs the adaptive filter so that it no longer sufficiently attenuates the echo. Unfortunately, algorithms having a high convergence rate usually diverge quickly in the presence of double-talk. To limit the divergence of the echo canceler (EC) during double-talk, the standard procedure is to use a level-based double-talk detector (DTD), e.g., Geigel's algorithm [37]. Whenever double-talk is detected, the step size of the adaptive filtering algorithm is set to zero, thus inhibiting the adaptation. Unfortunately, during the time required by the DTD to detect double-talk, the echo canceler often diverges. This is because a few undetected samples may perturb the echo path estimate considerably and the coefficients are then frozen in a poor state for at least the *hangover time*, which is the minimum time for which adaptation is inhibited after detection of double-talk.

Besides presenting the proportionate step-size technique, this chapter will also describe a very successful method that is used to decelerate the divergence of the algorithm caused by undetected double-talk, while maintaining good convergence. Our approach has its roots in the field of robust statistics and is based on introducing a scaled nonlinearity into the adaptive algorithm. The nonlinearity limits the impact of large disturbances on the coefficient setting, but it also reduces the convergence rate somewhat. The combination of the nonlinearity and the proportional step-size technique gives us robust versions of PNLMS, SR-PNLMS, and PAPA.

Another improvement to the overall performance of the adaptive filter, not described in this book, is to allow continuous variability of the global step size parameter (μ). See [137], [2], [24] for a thorough discussion of this approach.

This chapter is organized as follows: Notation and adaptive algorithms are presented in Sect. 2.2. A short description of the Geigel double-talk detector and a motivation for robust algorithms are given in Sect. 2.3. In Sect. 2.4, we derive an adaptive scale factor, present robust versions of NLMS, PNLMS, and SR-PNLMS, and generalize the robust PNLMS algorithm to a robust proportionate affine projection algorithm. Section 2.5 shows the computational complexity of the robust principle and compares the complexity of the different adaptive algorithms. Some aspects regarding the numerical dynamic range of the algorithms are also discussed. Simulation results are shown in Sect. 2.6 and a discussion and conclusions are given in Sect. 2.7.

2.2 Adaptive Proportionate Step-size Algorithms

Referring to Fig. 2.1, the following notation is used in all the derivations:

$$x(n) = \text{Far-end signal/speech,}$$
$$w(n) = \text{Ambient (background) noise,}$$
$$v(n) = \text{Near-end signal/speech (double-talk),}$$
$$y(n) = \text{Echo and ambient noise, possibly including}$$
$$\text{near-end signal,}$$
$$\mathbf{y}(n) = [y(n) \cdots y(n - M + 1)]^T , \text{ Vector of samples } y(n),$$
$$\mathbf{x}(n) = [x(n) \cdots x(n - L + 1)]^T , \text{ Excitation vector,}$$
$$\mathbf{X}(n) = [\mathbf{x}(n) \cdots \mathbf{x}(n - M + 1)] , \text{ Excitation matrix,}$$

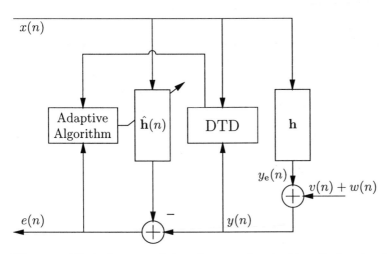

Fig. 2.1. Block diagram of the echo canceler, double-talk detector (DTD), and echo path

$$\mathbf{h} = [h_0 \; \cdots \; h_{L-1}]^T \; , \; \text{True echo path vector,}$$

$$\hat{\mathbf{h}}(n) = [h_0(n) \; \cdots \; h_{L-1}(n)]^T \; , \; \text{Estimated echo path vector,}$$

$$e(n) = y(n) - \hat{\mathbf{h}}^T(n-1)\mathbf{x}(n) \; , \; \text{Residual error,}$$

$$\mathbf{e}(n) = \mathbf{y}(n) - \mathbf{X}^T(n)\hat{\mathbf{h}}(n-1) \; , \; \text{Residual error vector,}$$

$$e_m(n) = m\text{th element of the vector } \mathbf{e}(n) \; .$$

Here, L is the adaptive filter length and M is the projection order of the affine projection algorithm.

The purpose of the adaptive filter is to estimate the echo path so that it can subtract a replica of the returned echo, $y_e(n) = \mathbf{h}^T\mathbf{x}(n)$; see Fig. 2.1. Traditionally, the NLMS algorithm has been the workhorse in echo canceler implementations. It therefore serves as a reference algorithm. The coefficient update equation is given by,

$$\hat{\mathbf{h}}(n) = \hat{\mathbf{h}}(n-1) + \frac{\mu}{\mathbf{x}^T(n)\mathbf{x}(n) + \delta}\mathbf{x}(n)e(n) \; , \tag{2.1}$$

where μ is the overall step-size parameter (relaxation parameter), and $\delta > 0$ is a regularization parameter which prevents division by zero and stabilizes the solution.

2.2.1 The PNLMS Algorithm

Network echo paths are usually sparse, which means that a large number of coefficients in the impulse response are effectively zero. Hence, it is important to identify the non-zero (active) coefficients. An algorithm that takes this property into account is the PNLMS algorithm which was proposed in

[38]. In this algorithm, an adaptive individual step size is assigned to each filter coefficient. The step sizes are calculated from the last estimate of the filter coefficients so that larger coefficients receive a larger step size, thus increasing the convergence rate of that coefficient. This has the effect that active coefficients are adjusted faster than non-active coefficients (i.e., small or zero coefficients). An advantage of this technique compared to other approaches to individual step-size algorithms, e.g., [124], [67], is that less *a priori* information is needed. Furthermore, the active regions of the echo path do not have to be detected explicitly. Another interesting idea for step-size control is presented in [64]. The basic PNLMS algorithm is described by the following equations:

$$\hat{\mathbf{h}}(n) = \hat{\mathbf{h}}(n-1) + \frac{\mu}{\mathbf{x}^T(n)\mathbf{G}(n)\mathbf{x}(n) + \delta}\mathbf{G}(n)\mathbf{x}(n)e(n) \,, \tag{2.2}$$

$$\mathbf{G}(n) = \mathrm{diag}\{g_0(n) \cdots g_{L-1}(n)\} \,. \tag{2.3}$$

Here $\mathbf{G}(n)$ is a diagonal matrix that adjusts the step sizes of the individual taps of the filter. The diagonal elements of $\mathbf{G}(n)$ are calculated as [38]

$$\gamma_{\min} = \varrho \max\{\delta_{\mathrm{p}}, |\hat{h}_0(n)|, \ldots, |\hat{h}_{L-1}(n)|\} \,, \tag{2.4a}$$

$$\gamma_l(n+1) = \max\{\gamma_{\min}, |\hat{h}_l(n)|\} \,, \ 0 \le l \le L-1 \,, \tag{2.4b}$$

$$g_l(n+1) = \frac{\gamma_l(n+1)}{\dfrac{1}{L}\displaystyle\sum_{i=0}^{L-1}\gamma_i(n)} \,, \ 0 \le l \le L-1 \,, \tag{2.4c}$$

where the parameters δ_{p} ($_\mathrm{p}$ denotes PNLMS) and ϱ are positive numbers with typical values $\delta_{\mathrm{p}} = 0.01$, $\varrho = 5/L$. The term $\gamma_{\min}$ prevents $\hat{h}_l(n)$ from stalling when it is much smaller than the largest coefficient and δ_{p} regularizes the update when all coefficients are zero at initialization.

A variant of this algorithm is called the PNLMS++ [54]. In this algorithm, for odd-numbered time steps the matrix $\mathbf{G}(n)$ is derived as above, while for even-numbered steps it is chosen to be the identity matrix,

$$\mathbf{G}(n) = \mathbf{I} \,, \tag{2.5}$$

which leads to an NLMS iteration. The alternation between NLMS and PNLMS iterations has advantages compared to using just the PNLMS technique, e.g., it makes the PNLMS++ algorithm much less sensitive to the assumption of a sparse impulse response without sacrificing performance.

2.2.2 A Signed Regressor PNLMS Algorithm

A variant of the LMS algorithm, the signed regressor LMS algorithm, can be found in [30]. In this algorithm, the regressor, i.e., the excitation vector $\mathbf{x}(n)$, is replaced by a vector containing only the signs of the input samples, $x(n-l)$, $l = 0, \ldots, L-1$. The rationale for this algorithm is that it leads to a less complex (in terms of multiplications) algorithm. One drawback is slower convergence due to the distortion the sign operator causes in the gradient estimation.

Let the function sign($\mathbf{a}$) operate on each element in the arbitrary vector $\mathbf{a}$, i.e., it returns a vector containing ± 1 depending on the sign of each element. The signed regressor LMS (SR-LMS) coefficient update is given by,

$$\hat{\mathbf{h}}(n) = \hat{\mathbf{h}}(n-1) + \mu\,\text{sign}\,[\mathbf{x}(\text{n})]e(n)\,. \qquad (2.6)$$

For a non-stationary signal such as speech, the SR-LMS is not properly normalized, i.e., it has a different convergence rate depending on the power of the excitation signal. To alleviate this problem we propose a normalized version of (2.6) where we divide the update by the sum of the magnitude of the input samples, giving

$$\hat{\mathbf{h}}(n) = \hat{\mathbf{h}}(n-1) + \frac{\mu\,\text{sign}\,[\mathbf{x}(n)]\,e(n)}{\mathbf{x}^T(n)\text{sign}\,[\mathbf{x}(n)] + \delta/\{\mathbf{x}^T(n)\text{sign}\,[\mathbf{x}(n)] + \delta_\text{r}\}}\,. \quad (2.7)$$

This normalization is reasonable since the gradient in (2.6) (sign$\,[\mathbf{x}(\text{n})]e(n) = $ sign$\,[\mathbf{x}(\text{n})]\{[\mathbf{h} - \mathbf{h}(n)]^T\mathbf{x}(n) + w(n)\}$) is proportional to the magnitude of $x(n)$ for $w(n)$ small. Furthermore, the regularization parameter δ is also scaled by $\sum_{i=0}^{L-1}|x(n-i)| + \delta_\text{r} = \mathbf{x}^T(n)\text{sign}\,[\mathbf{x}(n)] + \delta_\text{r}$, where δ_r inhibits division by zero in (2.7). With this scaling of δ, the effect of regularizing (2.7) is the same as regularizing with the same δ in (2.1).

We now propose to combine the properly normalized and regularized signed regressor algorithm with the proportionate step-size matrix (2.3). Our motivation for doing this is to remedy the reduced convergence rate that results from using the SR-LMS, compared to NLMS, and achieve a reduction of the complexity compared to PNLMS. We start with the basic signed regressor PNLMS (SR-PNLMS) algorithm,

$$\hat{\mathbf{h}}(n) = \hat{\mathbf{h}}(n-1)$$
$$+ \frac{\mu\,\mathbf{G}(n)\text{sign}\,[\mathbf{x}(n)]e(n)}{\mathbf{x}^T(n)\mathbf{G}(n)\text{sign}\,[\mathbf{x}(n)] + \delta/\{\mathbf{x}^T(n)\text{sign}\,[\mathbf{x}(n)] + \delta_\text{r}\}}\,. \qquad (2.8)$$

A "++" version in the same manner as for PNLMS++ can also be given for this algorithm by, e.g., alternating between (2.8) and (2.7).

Algorithm (2.8) still has a rather high numerical complexity because of the term, $\mathbf{x}^T(n)\mathbf{G}(n)\text{sign}\,[\mathbf{x}(n)]$, which consumes L multiplications. A more efficient normalization that works well for sparse echo paths is

$$\hat{\mathbf{h}}(n) = \hat{\mathbf{h}}(n-1)$$
$$+ \frac{\mu\,\mathbf{G}(n)\text{sign}\,[\mathbf{x}(n)]e(n)}{\theta g_{\text{max}}(n)\mathbf{x}^T(n)\text{sign}\,[\mathbf{x}(n)] + \delta/\{\mathbf{x}^T(n)\text{sign}\,[\mathbf{x}(n)] + \delta_\text{r}\}}\,, \qquad (2.9)$$

where

$$g_{\text{max}}(n) = \max_l\,[g_l(n)]\,, \qquad (2.10)$$

and θ is a constant which is adjusted to make (2.9) behave similar to (2.8). Typically, $\theta = 10/L$. For dispersive echo paths however, this algorithm may become unstable. Note that this technique for reducing complexity can be used in the PNLMS algorithm (2.2) as well.

2.2.3 A Proportionate Affine Projection Algorithm (PAPA)

It is natural to generalize the PNLMS algorithm to a proportionate affine projection algorithm. (Remember that NLMS is a member of the affine projection algorithm family [100], [106].) A proportionate affine projection algorithm (PAPA) is given by

$$\hat{\mathbf{h}}(n) = \hat{\mathbf{h}}(n-1) + \mu \mathbf{G}(n)\mathbf{X}(n) \left[\mathbf{X}^T(n)\mathbf{G}(n)\mathbf{X}(n) + \delta \mathbf{I}\right]^{-1} \mathbf{e}(n) , \quad (2.11)$$

where $\mathbf{G}(n)$ is as defined in (2.3) and (2.4). Let

$$\mathbf{\Upsilon}_{xx}^{-1}(n) = \left[\mathbf{X}^T(n)\mathbf{G}(n)\mathbf{X}(n) + \delta \mathbf{I}\right]^{-1} . \quad (2.12)$$

This matrix decorrelates the input data, $\mathbf{X}(n)$, and thus the convergence rate of the adaptive filter is increased. With $\mathbf{G}(n) = \mathbf{I}$, equation (2.11) reduces to the standard APA, first introduced in [106]. The regularization parameter in (2.11) was proposed in [82]. As evident, PAPA is obtained by combining APA with the proportional step size of PNLMS. Since the inclusion of $\mathbf{G}(n)$ in (2.12) requires M^2L multiplications per sample we can significantly reduce the computations by omitting it. According to our simulations, the effect on performance and stability is minimal. Furthermore, most of the computational procedures of the fast affine projection (FAP) algorithm [55], [125] can be incorporated in order to reduce the computational complexity of PAPA. Unfortunately, introducing an alternative coefficient vector, as done in [55], cannot be done in PAPA because $\mathbf{G}(n)$ varies from one iteration to the next, and so, destroys invariance of the product of the step-size matrix and the excitation vector.

2.3 The Geigel DTD

A double-talk detector (DTD) is used to stop adaptation during periods of simultaneous far-end and near-end speech. A simple and efficient way of detecting double-talk is to compare the magnitude of the far-end and near-end signals and declare double-talk if the near-end magnitude becomes larger than a value set by the far-end speech. A proven algorithm that has been in commercial use for many years is the Geigel DTD [37]. In this algorithm, double-talk is declared if

$$\max\{ |x(n)|, \ldots, |x(n-L+1)| \} < T|y(n)| , \quad (2.13)$$

where T is a threshold constant. The detector threshold T, is set to 2 if the hybrid attenuation is assumed to be 6 dB, and to $\sqrt{2}$ if the attenuation is assumed to be 3 dB. A so-called hangover time (hold time) t_{hold}, is also specified such that if double-talk is detected, then the adaptation is inhibited for this duration beyond the detected end of double-talk. Although this detector works fairly well, detection errors do occur, and these result in large amounts of divergence of the adapted filter coefficients, which in turn give rise to large amounts of uncancelled echo. This problem is illustrated in Fig. 2.2. Figure 2.2(a), (b) show examples of far-end speech and near-end speech (double-talk) respectively. Figure 2.2(c) shows the disturbance resulting from double-talk detection errors ($T = 2$). This disturbance, which in practice cannot be measured, is generated by gating the DTD's decision with the pure double-talk sequence of Fig. 2.2(b) (which in general is not available).

One way to model these large disturbances together with the background noise is to represent them by an outlier-contaminated stochastic process. Note that an outlier-contaminated model is not appropriate for the residual error in the absence of a DTD, Fig. 2.2(b), because without the DTD the residual consists of the long-lasting bursts of near-end speech. Therefore, the DTD is an essential component in the robust algorithms to be described in the next section.

2.4 The Robust Algorithms

The adaptive algorithms presented above can all be made robust to large burst disturbances by modification of the criteria on which these algorithms are based. In general, however, such modifications lower the convergence rate. The difficult problem of robust echo cancellation is to devise an algorithm that diverges slowly in response to double-talk, yet is able to rapidly track changes of the echo path when they occur. These two requirements are contradictory. The key to our solution to this problem is a combination of a DTD with traditional robust statistics methods using an adaptive scale variable, s.

Recall that the LMS is an iterative algorithm that adjusts the estimated impulse response so as to minimize the cost function, $E\{|e(n)|^2\}$, i.e., the mean-squared error. Each iteration updates the current estimate of $\mathbf{h}(n)$ by $\mu\mathbf{x}(n)e(n)$, which is a step in the direction of a stochastic approximation to the gradient of $E\{|e(n)|^2\}$. To make the algorithm insensitive to changes of the level of the input signal, $x(n)$, the proportionate factor μ is normalized, resulting in the NLMS algorithm. It is well known, [117], that other gradient algorithms can be derived by changing the cost function to

$$J = E\left\{\rho\left(\frac{|e(n)|}{s}\right)\right\} , \qquad (2.14)$$

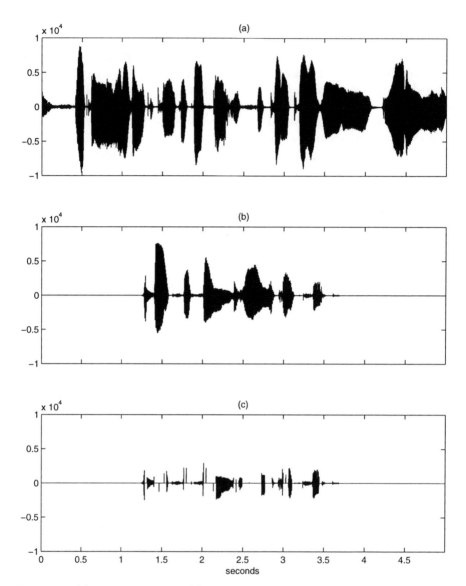

Fig. 2.2. (a) Far-end speech. (b) Near-end speech, i.e. double-talk. (c) Near-end speech gated with the Geigel DTD's decision. This is the disturbance that actually enters the adaptive algorithm. Average far-end to near-end ratio: 6 dB (1.125−3.625 s), $T = 2$

where $\rho(\cdot)$ is any symmetric function with a monotonically non-decreasing derivative (with respect to its argument)[1]. The variable s is an important

---

[1] More generally as discussed in [68], one can use (maximum likelihood) M-estimators which are defined as $J = \sum_n \rho(|e(n)|/s)$. The choice used in (2.14)

scale factor. The resulting algorithm, analogous to a relaxed Newton method, is

$$\hat{\mathbf{h}}(n) = \hat{\mathbf{h}}(n-1) - \mu \left[\nabla^2 J\right]^{-1} \nabla J \,, \tag{2.15}$$

where ∇ is the gradient and $\nabla^2 = \nabla\nabla^T$ is the Hessian, both with respect to $\hat{\mathbf{h}}$. The algorithm can be made robust by a proper choice of $\rho(\cdot)$, which should be chosen such that $\lim_{|e(n)|\to\infty} |\nabla\rho(|e(n)|/s)| < \infty$. Following suggestions in [68], we choose

$$\rho(|z|) = \begin{cases} \dfrac{|z|^2}{2} \,, & |z| \le k_0 \\ k_0|z| - \dfrac{k_0^2}{2} \,, & |z| > k_0 \end{cases} \tag{2.16}$$

and calculate

$$\nabla J = E\left\{ -\text{sign}\left[e(n)\right] \psi \left[\frac{|e(n)|}{s}\right] \frac{\mathbf{x}(n)}{s} \right\} \,, \tag{2.17}$$

where $\psi(|z|) = \dfrac{d\rho(|z|)}{d|z|}$ is a limiter,

$$\psi\left(\frac{|e(n)|}{s}\right) = \min\left\{ \frac{|e(n)|}{s}, k_0 \right\} \,. \tag{2.18}$$

Figure 2.3 illustrates the $\psi(\cdot)$ function. The effect of the scale factor s, and the manner in which it is adapted are discussed in Sect. 2.4.1. When the Newton method is utilized to derive the iterative algorithm, the inverse Hessian of the minimization criterion (2.14) $\left[(\nabla^2 J)^{-1}\right]$ should be used as a step-size matrix. If we assume that the excitation vector $\mathbf{x}(n)$ is independent of the error $e(n)$ (which is true upon convergence) we find the Hessian straightforwardly as,

$$\begin{aligned} \nabla^2 J &= \nabla\left[E\left\{ -\text{sign}\left[e(n)\right] \psi \left(\frac{|e(n)|}{s}\right) \frac{\mathbf{x}(n)}{s} \right\} \right]^T \\ &= E\left\{ -\left[\nabla\{\text{sign}\left[e(n)\right]\}\psi\left(\frac{|e(n)|}{s}\right) \right.\right. \\ &\quad\left.\left. + \text{sign}\left[e(n)\right] \nabla\{\psi\left(\frac{|e(n)|}{s}\right)\} \right] \frac{\mathbf{x}^T(n)}{s} \right\} \\ &= E\left\{ -\left[\mathbf{0}^T - \text{sign}^2\left[e(n)\right] \psi'\left(\frac{|e(n)|}{s}\right) \frac{\mathbf{x}(n)}{s} \right] \frac{\mathbf{x}^T(n)}{s} \right\} \\ &= E\left\{ \psi'\left(\frac{|e(n)|}{s}\right) \frac{\mathbf{x}(n)\mathbf{x}^T(n)}{s^2} \right\} = \frac{c}{s^2}\mathbf{R}_{xx} \,, \end{aligned} \tag{2.19}$$

i.e., a constant (c) times the correlation matrix $(\mathbf{R}_{xx})$ of the input signal where $c = E\{\psi'(|e(n)|/s)\}$. This value comes from our choice of criterion,

_____

makes the derivation of the iterative algorithm more consistent with the derivation of the LMS algorithm.

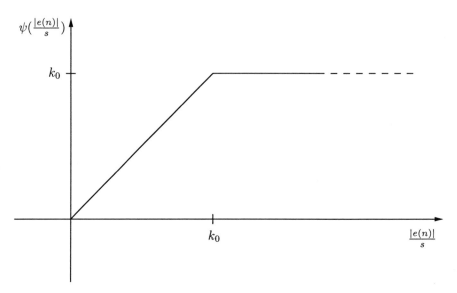

Fig. 2.3. Non-linear function, $\psi(\cdot)$, used for the robust algorithms (Huber function)

(2.18), whose derivative is either 0 or 1 depending on the magnitude of $e(n)/s$. In order to reduce complexity, we approximate the Hessian as a diagonal matrix,

$$\widehat{\nabla^2 J} = c\frac{\mathbf{x}^T(n)\mathbf{x}(n)}{s^2}\mathbf{I} \,, \tag{2.20}$$

and use a stochastic approximation of the gradient (2.17) in the update equation (2.15). This leads to the robust NLMS algorithm

$$\hat{\mathbf{h}}(n) = \hat{\mathbf{h}}(n-1) + \frac{\mu\mathbf{x}(n)}{\mathbf{x}^T(n)\mathbf{x}(n) + \delta}\psi\left(\frac{|e(n)|}{s}\right)\text{sign}\left[e(n)\right]s \,. \tag{2.21}$$

In this algorithm, we have also introduced the previously discussed regularization parameter δ. The constant c, which generally is very close to one, is now included in the step size μ.

The PNLMS algorithm given in (2.2) can be made robust in an analogous manner, yielding the update equation shown in Table 2.1. This table also shows the double-talk detector. Tables 2.2, 2.3, and 2.4 show the robust SR-PNLMS, PAPA, and a fast PAPA of order 2. The nonlinear function in all robust algorithms requires a scale factor $[s(n)]$ to be estimated. One way to estimate this factor will be discussed next.

2.4.1 Estimating the Scale Factor

The estimate of the scale factor s, should reflect the background noise level at the near end, be robust to short-burst disturbances (due to double-talk),

---

Table 2.1 The robust PNLMS algorithm and the Geigel double-talk detector

---

Step-size calculation

$$\gamma_l(n+1) = \max\{\varrho \max\{\delta_{\mathrm{p}}, |h_0(n)|, \ldots, |h_{L-1}(n)|\}, |h_l(n)|\}, \; 0 \le l \le L-1$$

$$g_l(n+1) = \frac{\gamma_l(n+1)}{\frac{1}{L}\sum_{i=0}^{L-1}\gamma_i(n)}, \; 0 \le l \le L-1$$

$$\mathbf{G}(n) = \mathrm{diag}\{g_0(n) \cdots g_{L-1}(n)\}$$

Geigel DTD

$$\max\{|x(n)| \cdots |x(n-L+1)|\} < T|y(n)| \; ; \quad \text{double-talk, } \mu = 0$$
$$\max\{|x(n)| \cdots |x(n-L+1)|\} \ge T|y(n)| \; ; \quad \text{no double-talk, } \mu = \mu_0$$
$$0 \le \mu_0 \le 2$$

Filtering

$$e(n) = y(n) - \hat{\mathbf{h}}^T(n-1)\mathbf{x}(n)$$

$$\psi\left[\frac{|e(n)|}{s(n)}\right] = \min\left[\frac{|e(n)|}{s(n)}, k_0\right]$$

$$\hat{\mathbf{h}}(n) = \hat{\mathbf{h}}(n-1) + \frac{\mu\mathbf{G}(n)\mathbf{x}(n)}{\mathbf{x}^T(n)\mathbf{G}(n)\mathbf{x}(n) + \delta}\psi\left[\frac{|e(n)|}{s(n)}\right]\mathrm{sign}\left[e(n)\right]s(n)$$

Table 2.2 The robust SR-PNLMS algorithm

Normalization update

$$\tilde{\delta}(n) = \tilde{\delta}(n-1) + |x(n)| - |x(n-L)| \; , \quad \tilde{\delta}(-1) = \delta_{\mathrm{r}}$$

Filtering

$$e(n) = y(n) - \hat{\mathbf{h}}^T(n-1)\mathbf{x}(n)$$

$$\psi\left[\frac{|e(n)|}{s(n)}\right] = \min\left[\frac{|e(n)|}{s(n)}, k_0\right]$$

$$\hat{\mathbf{h}}(n) = \hat{\mathbf{h}}(n-1)$$

$$+ \frac{\mu\mathbf{G}(n)\mathrm{sign}\{\mathbf{x}(n)\}}{\mathbf{x}^{\tilde{T}}(n)\mathbf{G}(n)\mathrm{sign}\{\mathbf{x}(n)\} + \delta/\tilde{\delta}(n)}\psi\left[\frac{|e(n)|}{s(n)}\right]\mathrm{sign}\left[e(n)\right]s(n)$$

---

and track long-term changes of the residual error (due to echo path changes). To fulfill these requirements, we have chosen the scale factor estimate as

$$s(n+1) = \lambda_{\mathrm{s}}s(n) + \frac{1-\lambda_{\mathrm{s}}}{\beta}\psi\left[\frac{|e(n)|}{s(n)}\right]s(n) \; , \tag{2.22}$$

Table 2.3 The robust PAPA algorithm. The symbol $\odot$ denotes elementwise (Hadamard) multiplications

Filtering

$$e(n) = y(n) - \mathbf{X}^T(n)\hat{\mathbf{h}}(n-1)$$

$$\mathbf{\Upsilon}_{xx}^{-1}(n) = \left[\mathbf{X}^T(n)\mathbf{G}(n)\mathbf{X}(n) + \delta\mathbf{I}\right]^{-1}$$

$$\psi_m\left[e_m(n)\right] = \min\left(\frac{|e_m(n)|}{s(n)}, k_0\right), \ m = 0, \ldots M-1$$

$$\hat{\mathbf{h}}(n) = \hat{\mathbf{h}}(n-1) + \mu\mathbf{G}(n)\mathbf{X}(n)\mathbf{\Upsilon}_{xx}^{-1}(n)\left\{\psi\left[\mathbf{e}(n)\right] \odot \text{sign}\left[\mathbf{e}(n)\right]\right\}s(n)$$

Table 2.4 A fast implementation of the robust PAPA algorithm of order 2. The vector $\mathbf{\Upsilon}_{xx}^{-1}(n)\mathbf{e}(n)$ can be simply calculated by directly solving the 2 by 2 system of equations

Projection matrix update

$$r_{22}(n) = r_{11}(n-1)$$

$$r_{11}(n) = r_{11}(n-1) + x^2(n) - x^2(n-L)$$

$$r_{12}(n) = r_{12}(n-1) + x(n)x(n-1) - x(n-L)x(n-L-1)$$

$$\mathbf{\Upsilon}_{xx}(n) = \begin{bmatrix} r_{11}(n) & r_{12}(n) \\ r_{12}(n) & r_{22}(n) \end{bmatrix}$$

Filtering

$$e(n) = d(n) - \hat{\mathbf{h}}^T(n-1)\mathbf{x}(n),$$

$$\psi(|e(n)|) = \min\{|e(n)|, k_0 s(n)\}$$

$$\mathbf{e}(n) = \left[\psi(|e(n)|)\text{sign}\left[e(n)\right] \ (1-\mu)\mathbf{e}_1^T(n-1)\right]^T$$

$$\hat{\mathbf{h}}(n) = \hat{\mathbf{h}}(n-1) + \mu\mathbf{G}(n)\mathbf{X}(n)\mathbf{\Upsilon}_{xx}^{-1}(n)\mathbf{e}(n)$$

---

where β is a normalization constant (defined later in Sect. 2.4.2). Our choice of this method of estimating s, which is very simple to implement, is also justified in Sect. 2.4.2. With this choice, the current estimate of s is governed by the level of the error signal in the immediate past over a time interval roughly equal to $1/(1-\lambda_s)$ samples where the practical range of λ_s is $[0.99, 1)$ and the range of β [which depends on k_0 in (2.18)] is approximately $[0.60, 0.74]$. The initial value of the scale factor can be chosen as $s(0) = \sigma_x$, where σ_x is set approximately to the average speech level in voice telephone networks, $\approx -18\,\text{dBm0}$, which corresponds to the linear value ≈ 2900 in a 16-bit representation. (In telecommunications, $3\,\text{dBm0}$ corresponds to the maximum

Table 2.5 Scale factor updating controlled by the double-talk decision

if (no double-talk detected)

$$s(n + 1) = \lambda_s s(n) + \frac{(1 - \lambda_s)}{\beta} \psi \left[\frac{|e(n)|}{s(n)} \right] s(n)$$

else

$$s(n + 1) = \lambda_s s(n) + (1 - \lambda_s) s_{\min}$$

end

amplitude level in the system.) When the algorithm has not yet converged, s is large. Hence, the limiter is in its linear portion and therefore the robust algorithm behaves like the conventional NLMS or PNLMS algorithm. When double-talk occurs, the error is determined by the limiter and by the scale of the error signal, which depends on the recent past of the error signal before the double-talk occurs. Thus, the divergence rate is reduced over intervals of double-talk less than about $1/(1 - \lambda_s)$ samples. This gives ample time for the DTD to act. If there is a sudden system change, the algorithm will not track immediately. However, as the scale estimator starts to adapt to the larger error signal, the nonlinearity is scaled up and the convergence rate accelerates. The trade-off between robustness and the tracking rate of the adaptive algorithm is thus governed by the tracking rate of the scale estimator, which is controlled by a single parameter, λ_s. As with the Geigel DTD, it is useful to introduce a hangover time for control of scale updating. When the DTD detects double-talk, adaptation of $s(n)$ should be inhibited for a specific time, preferably as long as the DTD hangover interval, t_{hold}. Moreover, double-talk may have already affected $s(n)$ (biased it towards an unacceptably large value) by the time it is detected. We therefore let the scale estimate decay towards a minimum value $s_{\min}$. This procedure, shown in Table 2.5, ensures that any disturbance of the scale estimate slowly decays when double-talk has been detected.

2.4.2 Detailed Derivation of the Scale Estimate

A scale factor can be found by defining an implicit function, denoted by J_s, which is a weighted sum of function values $\chi(\cdot)$, given by [68],

$$J_s(n) = \sum_{l=0}^{n} \lambda_1^{n-l} \chi \left[\frac{|e(l)|}{s} \right] = 0 \,, \tag{2.23}$$

where s is the scale factor and λ_1 is a forgetting factor. For a bounded $\chi(\cdot)$ it is obvious that s can be chosen such that (2.23) is fulfilled, i.e., s scales the value of J_s. We have chosen $\chi(\cdot)$ as,

$$\chi(\cdot) = \psi(\cdot) - \beta \, , \tag{2.24}$$

where $\psi(\cdot)$ is given in (2.18) and β is a positive constant. The reason for using $\psi(\cdot)$ here as well as in Table 2.1, is purely for reducing computational complexity. Other bounded even functions would also suffice since the choice of $\chi(\cdot)$ turns out to be not very critical. For normalization, β is chosen such that for a zero-mean, unit-variance Gaussian process z,

$$E\{\chi(z)\} = 0 \, .$$

Using (2.18) for $\psi(\cdot)$ gives

$$\begin{aligned}
\beta &= \frac{2}{\sqrt{2\pi}} \int_0^\infty \psi\{z\} e^{-\frac{1}{2}z^2} dz \\
&= \sqrt{\frac{2}{\pi}} (1 - e^{-\frac{1}{2}k_0^2}) + k_0 \mathrm{erfc}\left(\frac{k_0}{\sqrt{2}}\right) ,
\end{aligned} \tag{2.25}$$

where

$$\mathrm{erfc}(x) = \frac{2}{\sqrt{\pi}} \int_x^\infty e^{-t^2} dt \, . \tag{2.26}$$

Upon convergence, the scale estimate $s = \sigma_v$ for Gaussian noise, where σ_v is the desired standard deviation of the (near-end) ambient noise.

A recursive scale estimator can then be derived with a Newton technique. Let

$$s(n+1) = s(n) - [\nabla_s J_s]^{-1} J_s \, . \tag{2.27}$$

The gradient of (2.23) is

$$\nabla_s J_s(n) = \sum_{l=0}^n -\lambda_1^{n-l} \frac{|e(l)|}{s^2} \chi'\left[\frac{|e(l)|}{s}\right] \, . \tag{2.28}$$

A simple update equation for the scale (2.27) is found by introducing the following approximations,

$$\hat{J}_s(n) = \chi\left[\frac{|e(n)|}{s(n)}\right] , \tag{2.29}$$

$$\widehat{\nabla_s J_s}(n) = E\{\nabla_s J_s(n)\} \, . \tag{2.30}$$

Assuming the gradient $\nabla_s J_s$ to be stationary and that $s(n)$ converges to the background standard deviation of the noise (Gaussian), we find,

$$E\{\nabla_s J_s(n)\} = \sum_{l=0}^{n} -\lambda_1^{n-l} E\left\{\frac{|e(l)|}{s^2}\chi'\left[\frac{|e(l)|}{s}\right]\right\}$$

$$= \sum_{l=0}^{n} -\lambda_1^{n-l}\frac{1}{s} E\left\{\frac{|e(l)|}{s}\chi'\left[\frac{|e(l)|}{s}\right]\right\}$$

$$= \sum_{l=0}^{n} -\lambda_1^{n-l}\frac{1}{s}\frac{2}{\sqrt{2\pi}}\int_0^{k_0} z e^{-\frac{1}{2}z^2}\,dz$$

$$= \sum_{l=0}^{n} -\lambda_1^{n-l}\frac{1}{s}\sqrt{\frac{2}{\pi}}(1 - e^{-\frac{1}{2}k_0^2}) \xrightarrow[n\to\infty]{} -\frac{1}{s}\frac{\vartheta}{1-\lambda_1}, \quad (2.31)$$

where

$$\vartheta = \sqrt{\frac{2}{\pi}}(1 - e^{-\frac{1}{2}k_0^2}). \tag{2.32}$$

Combining (2.24), (2.27), (2.29), (2.30), and (2.31) leads to

$$s(n+1) = s(n) + \frac{(1-\lambda_1)s(n)}{\vartheta}\left\{\psi\left[\frac{|e(n)|}{s(n)}\right] - \beta\right\}$$

$$= \lambda_s s(n) + \frac{(1-\lambda_s)}{\beta}s(n)\psi\left[\frac{|e(n)|}{s(n)}\right], \tag{2.33}$$

where

$$\lambda_s = 1 - \frac{\beta}{\vartheta}(1-\lambda_1). \tag{2.34}$$

2.5 Complexity Comparison

It is interesting to compare the complexity of the presented algorithms. We first look at the complexity of the scaled nonlinearity. Redefine the calculations as,

$$\psi(|e(n)|) = \min\{|e(n)|, s(n)\}, \tag{2.35}$$

$$s(n+1) = \lambda_s s(n) + \frac{k_0(1-\lambda_s)}{\beta}\psi(|e(n)|), \tag{2.36}$$

which leads to the complexity in multiplications per sample shown in Table 2.6.

Table 2.6 Computational requirements for the scaled nonlinearity

Equation	Multiplications	Memory loc.
(2.35)	0	2
(2.36)	2	2
Total	2	4

Table 2.7 Approximative computational requirements for various algorithms including the scaled nonlinearity

Algorithm	Multiplications	Divisions
NLMS	$2L + 4$	1
PNLMS	$4L + 4$	1
SR-PNLMS (2.8)	$3L + 5$	2
SR-PNLMS (2.9)	$2L + 8$	2
PAPA (order $M > 2$)	$(M + 2)L + 18M$	5
PAPA ($M = 2$)	$4L + 12$	3

Looking at the NLMS (2.1), PNLMS (2.2), and modified SR-PNLMS (2.9) algorithms, we find that they require L multiplications for calculating the residual error $e(n)$. The normalization factor of NLMS can be recursively calculated using 2 multiplications and the gradient needs another L multiplications. The total for the basic NLMS is thus $2L + 2$ multiplications. Because of the individual step-size matrix $\mathbf{G}(n)$, PNLMS requires another $2L$ multiplications for the gradient calculation and normalization compared to NLMS. The SR-PNLMS algorithm, however, needs only L extra multiplications for the gradient calculation. Table 2.7 summarizes the complexity requirements for the different algorithms, including multiplications for the nonlinearity of the robust NLMS, PNLMS, SR-PNLMS and PAPA algorithms. The difference in complexity between PNLMS and PAPA (order 2) is modest.

Another important factor, when implementing an adaptive algorithm in fixed-point arithmetic, is the numerical range of the calculations. The maximum values of involved calculations can serve as measures of the numerical range of the algorithm. The step-size element $g_{\max}$ can be upper-bounded by L. However, this is a very unlikely value and will practically never occur. The following upper bounds on the intermediate calculations made in (2.1), (2.2), and (2.9) are useful:

$$|\mathbf{x}^T(n)\mathbf{x}(n)| \leq L|x_{\max}|^2 , \tag{2.37}$$

$$|\mathbf{x}^T(n)\text{sign}\{\mathbf{x}(n)\}| \leq L|x_{\max}| , \tag{2.38}$$

$$|\mathbf{x}^T(n)\mathbf{G}(n)\mathbf{x}(n)| \leq g_{\max}L|x_{\max}|^2 , \quad 0 < g_{\max} \leq L , \tag{2.39}$$

$$|\mathbf{x}^T(n)\mathbf{G}(n)\text{sign}\{\mathbf{x}(n)\}| \leq g_{\max}L|x_{\max}| , \quad 0 < g_{\max} \leq L , \tag{2.40}$$

$$||\mathbf{G}(n)\mathbf{x}(n)||_\infty \leq g_{\max}|x_{\max}| , \quad 0 < g_{\max} \leq L , \tag{2.41}$$

$$||\mathbf{G}(n)\text{sign}\{\mathbf{x}(n)\}||_\infty \leq g_{\max} , \quad 0 < g_{\max} \leq L . \tag{2.42}$$

Equations $(2.37)-(2.40)$ are computational results stored in the accumulator, and therefore they determine its required size. Equations (2.41) and (2.42) show the maximum dynamic range of calculations in the multiplier. Table 2.8 summarizes the numerical dynamic range of calculations for the previously presented algorithms. Comparing the maximum values required to be represented in the intermediate calculations of the algorithm, we find that the

Table 2.8 Maximum values in intermediate calculations required for NLMS, PNLMS, SR-PNLMS, and PAPA $(M = 2)$ algorithms

Algorithm	Multiplier	Accumulator
NLMS	$\lvert x_{\max}\rvert^2$	$L\lvert x_{\max}\rvert^2$
PNLMS	$L\lvert x_{\max}\rvert^2$	$L^2\lvert x_{\max}\rvert^2$
SR-PNLMS (2.8), (2.9)	$L\lvert x_{\max}\rvert$	$L^2\lvert x_{\max}\rvert$
PAPA $(M = 2)$	$L\lvert x_{\max}\rvert^2$	$L^2\lvert x_{\max}\rvert^2$

dynamic range of the calculations in the SR-PNLMS algorithm is roughly proportional to the square root of the corresponding calculations in the PNLMS algorithm. This property of SR-PNLMS is desirable from a numerical point of view.

2.6 Simulations

A full comparison of every aspect of the robust versus the non-robust algorithms is beyond the scope of this chapter. To reduce the number of simulations we state the following general properties of the algorithms:

- A robust version of an algorithm converges slower than its non-robust version. This means that the robust NLMS is slower than standard NLMS etc.
- Different algorithms are affected differently by the scaled nonlinearity, e.g., the robust NLMS is significantly slower than NLMS while the robust PNLMS is almost as fast as PNLMS.
- When double-talk occurs, the divergence rates of the non-robust, NLMS, PNLMS, and PAPA algorithms are approximately equal. This was shown in [54] for NLMS and PNLMS(++).

Comparisons between non-robust NLMS and PNLMS(++) with respect to convergence rate can be found in [38], [54]. Because of the above facts, we choose to compare the robust PAPA and PNLMS algorithms with the standard non-robust NLMS which has been the preferred algorithm in commercial echo cancelers. (We do not consider robust NLMS here because it converges even slower than non-robust NLMS and therefore is not very useful.) Further, we compare the robust SR-PNLMS algorithm with the (non-robust) PNLMS algorithm. The purpose of the simulations in this section is to illustrate the performance of robust algorithms during double-talk and echo path changes.

Important factors that affect the results are, e.g., type of excitation signal, far-end to double-talk ratio, and hybrid attenuation. Therefore, experiments using speech as well as the composite source signal (explained below) as excitation signal are shown. We choose the far-end to double-talk ratio approximately equal to the echo path attenuation assumption in the Geigel DTD. This can be regarded as a worst case. Experiments are performed with three different hybrid attenuations. All algorithms incorporate the Geigel DTD (Sect. 2.2) in which the settings (T, t_{hold}) are chosen the same as commonly used in commercial hardware. These settings have been found to be "optimal" in practice so that the DTD operates well for all different signal situations that may occur in a telephone network.

With a projection order of 2 in the PAPA algorithm, an AR(1) input process can be perfectly whitened and maximum improvement of the convergence rate is achieved. Speech, however, is not an AR(1) process but can be fairly well modeled by an AR(8) process. Choosing the order as 2 is a compromise between complexity and performance. Moreover, the improvement in performance is quite small when M is increased beyond 2, [55], [24].

2.6.1 Results with Speech as Excitation Signal

The parameter settings chosen for the following simulations are:

- All speech signal are in the range $[-32768, 32767]$.
- $\sigma_x = 1.9 \cdot 10^3$, echo-to-ambient-noise ratio ENR $= \sigma_{y_e}^2/\sigma_w^2$, is set to 8000 (39 dB).
- Average far-end to double-talk ratio is 6 dB.
- $\mu = 0.2$, $L = 512$ (64 ms), $\delta = 2 \cdot 10^5$ (NLMS, PNLMS, SR-PNLMS), $\delta = 1 \cdot 10^6$ (PAPA), $\delta_{\text{p}} = 0.01$, $\varrho = 0.01$.
- Hybrid attenuation: 20 dB.
- Geigel detector assumes 6 dB attenuation $(T = 2)$, $t_{\text{hold}} = 30$ ms (240 samples at 8 kHz sampling frequency).
- Parameters for the robust algorithm: $(\lambda_s, k_0) = (0.997, 1.1)$. This choice results in $\beta \approx 0.60665$.
- $\hat{\mathbf{h}}(0) = \mathbf{0}$, $s(0) = 1000$.

To inhibit poor behavior in low-noise situations, the scale estimate in (2.22), $s(n)$, is never allowed to become lower than 2. All algorithms are tuned to achieve the same asymptotic minimum mean-squared error in order to fairly compare convergence rate. Performance is measured by means of the normalized misalignment defined as,

$$\frac{\|\mathbf{h} - \hat{\mathbf{h}}(n)\|^2}{\|\mathbf{h}\|^2} . \tag{2.43}$$

The impulse response and corresponding frequency response magnitude of the hybrid are shown in Fig. 2.4(a), (b). (The responses in Fig. 2.4(c)−(f) are used in the next simulation.)

Figure 2.2(a), (b) shows the far-end and near-end (double-talk) signals. Figure 2.5(a) shows the misalignment of the three algorithms during double-talk. Initial convergence rates of PNLMS and PAPA are clearly superior to that of NLMS. While the non-robust NLMS (with Geigel detector) diverges

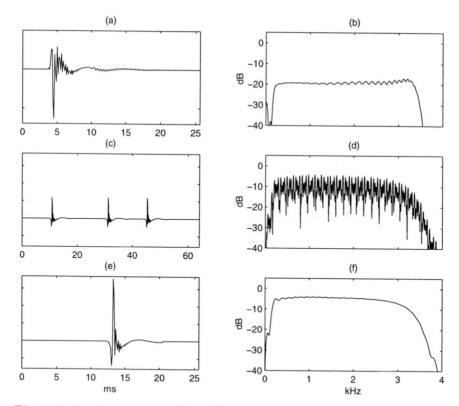

Fig. 2.4. Impulse responses (a), (c), (e), and frequency response magnitudes (b), (d), (f), of the hybrids in the simulations. Hybrid attenuation: 20 dB (a), (b); 11 dB (c), (d); 6 dB (e), (f)

to a misalignment of +5 dB, the robust algorithms are much less affected and never perform worse than −10 dB misalignment during double-talk. The slow reconvergence of NLMS and robust PNLMS after the double-talk sequence is caused by poor excitation of the far-end signal, i.e., this segment is highly correlated. Figure 2.5(b) shows the behavior after an abrupt system change, where the impulse response is shifted 200 samples at 1 second (near-end speech is *not* present in this simulation). The robust algorithms also outperform NLMS in this case.

Double-talk behaviour for the robust SR-PNLMS algorithm is shown in Fig. 2.6(a). We find that the robust SR-PNLMS performs far better than the non-robust PNLMS algorithm when exposed to double-talk. Figure 2.6(b) shows the re-convergence of the algorithms after the echo path change occurring at 1 s (near-end speech is *not* present in this simulation). The convergence rate of the robust SR-PNLMS is only slightly slower than that of PNLMS.

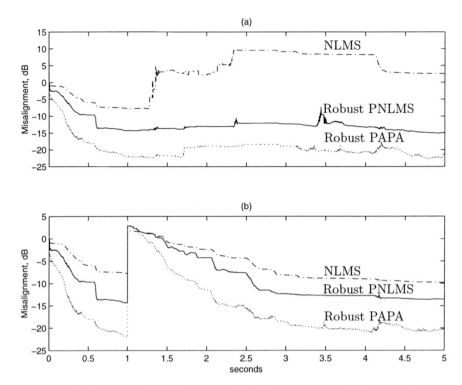

Fig. 2.5. Performance of NLMS, robust PNLMS, and robust PAPA algorithms for far-end and near-end (double-talk) speech in Fig. 2.2(a) and (b), respectively, and hybrid response in Fig. 2.4(a). (a) Misalignment during double-talk. (b) Misalignment after abrupt hybrid changes. Parameter values: $\lambda_s = 0.997$, $k_0 = 1.1$

2.6.2 Results with the Composite Source Signal as Excitation Signal

The ITU-T G.168 standard [69] recommends certain test procedures for evaluating the performance of echo cancelers. The test signal described in G.168 is a composite source signal (CSS) that has properties similar to those of speech with both voiced and unvoiced sequences as well as pauses. This section presents results from test situations evaluating the performance during double-talk, the so-called test 3B. The (normalized) mean-square error is defined as the performance index and is given by,

$$\text{MSE} = \frac{\text{LPF}\{[e(n) - v(n)]^2\}}{\text{LPF}\{[y(n) - v(n)]^2\}} \, , \tag{2.44}$$

where $\text{LPF}\{\cdot\}$ denotes a lowpass filter operation. In this case, the LPF has a single pole at 0.999 (time constant of $1/8$ s at 8 kHz sampling rate). This choice does not significantly affect the convergence rate in the figures.

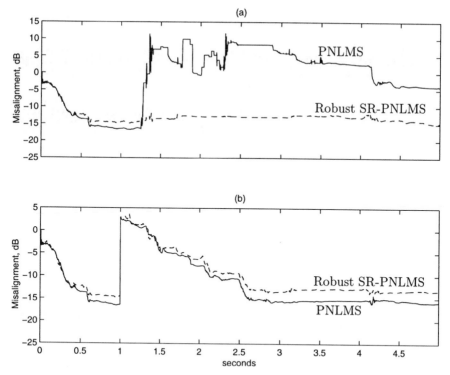

Fig. 2.6. Performance of PNLMS, and robust SR-PNLMS algorithms. (a) Misalignment during double-talk. (b) Misalignment after abrupt hybrid changes. Other conditions same as in Fig. 2.5

Test 3B evaluates the performance of the echo canceler for a high-level double-talk sequence. The double-talk level in this case is about the same as that of the far-end signal. Thus, a fixed threshold Geigel DTD assuming some nominal hybrid attenuation is able to detect this double-talk. False alarms and failures to detect double-talk are influenced by the chosen threshold in the DTD and the attenuation of the hybrid. Results from two hybrid attenuations are therefore evaluated. What differs in parameter and initial value settings from the previous simulation are:

- $\sigma_x = 1.3 \cdot 10^3$, ENR ≈ 37 dB (echo-to-ambient-noise ratio, where the noise is due to PCM quantization).
- Hybrid attenuation: 6 dB or 11 dB.
- Geigel detector assumes 3 dB attenuation ($T = \sqrt{2}$), $t_{\text{hold}} = 30$ ms.

The tests are made using a sparse echo path with 6 dB attenuation and a multireflection echo path with 11 dB attenuation. These are shown in Fig. 2.4(c)−(f).

(a)

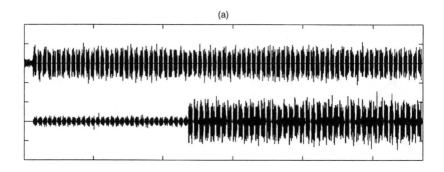

(b)

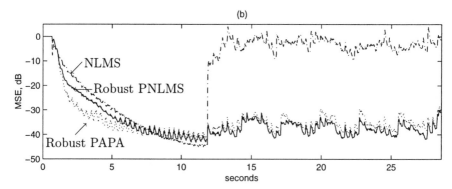

Fig. 2.7. Performance of NLMS, robust PAPA, and robust PNLMS algorithms for CSS input during double-talk using a multi-reflection hybrid with 11 dB attenuation [Fig. 2.4(c), (d)]. (a) Far-end (upper) and near-end (lower) signals; average far-end to double-talk ratio: 0 dB (12–31 s). (b) MSE. Parameter values: $\lambda_s = 0.9975$, $k_0 = 1.5$; other conditions same as in Fig. 2.5

Far-end and near-end signals used in test 3B are shown in Figs. 2.7(a) and 2.8(a) for the 11 dB and 6 dB hybrid respectively. Double-talk starts after about 12 seconds. Mean-square errors of the three algorithms are shown in Figs. 2.7(b) and 2.8(b). The latter case with a 6 dB hybrid is considered to be very difficult in practice. The two robust algorithms handle double-talk without severe degradation of MSE, while the non-robust NLMS, despite the Geigel detector, diverges as much as 30 dB. This divergence occurs when the DTD fails to detect the double-talk, and in the first case [Fig. 2.7(b)] as few as three samples drive the hybrid estimate far from optimum. This is the typical behaviour of a sample-by-sample update algorithm combined with a double-talk detector that makes a decision using the history of data up to the present sample (i.e., without "look-ahead"). In general, the lengths of undetected bursts in these simulations range from a few samples up to a couple of hundred samples. Moreover, a change of the threshold T, to a 6 dB assumption, would reduce the divergence of NLMS by only about 3–5 dB,

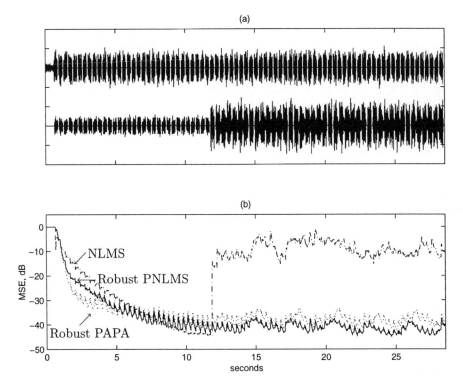

Fig. 2.8. Performance of NLMS, robust PAPA, and robust PNLMS algorithms for CSS input during double-talk using a sparse hybrid with 6 dB attenuation [Fig. 2.4(e), (f)]. (a) Far-end (upper) and near-end (lower) signals. (b) MSE. Other conditions same as in Fig. 2.7

while the false alarms would increase and slow down the convergence rate significantly, especially for a 6 dB hybrid. The performance of the robust SR-PNLMS algorithm is shown in Fig. 2.9. The loss incurred by using the sign of the regressor is very small when compared to the PNLMS algorithm.

2.7 Conclusions

A scaled nonlinearity combined with the Geigel DTD increases the robustness of echo cancelers. The scaled nonlinearity operates in the same manner as varying the step size (μ). That is, bounding the error signal can be interpreted as a reduction of the step-size parameter. What differentiates the robust approach is that while traditional variable step-size methods, [137], [2], try to detect periods of double-talk and then take action, the robust technique uses the signal in the absence of double-talk in order to be prepared when it occurs. Due to this fact, the robust technique is faster and more efficient.

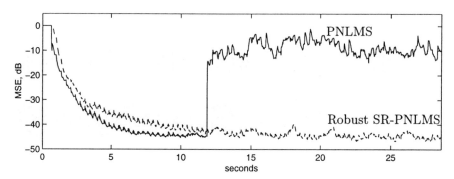

Fig. 2.9. Performance of the SR-PNLMS algorithm for CSS input during double-talk using a sparse hybrid with 6 dB attenuation [Fig. 2.4(e), (f)]. Other conditions same as in Fig. 2.7

Another major advantage is that only a few instructions and little memory are required to implement the robust principle.

The complexity of the PNLMS and PAPA algorithms is on the order of $4L$ multiplications per sample, which is about twice that of NLMS. However, the convergence rate of PNLMS and PAPA is considerably higher. It is shown in this chapter that the robust version of PNLMS and PAPA converge faster than NLMS and perform significantly better: up to 30 dB higher echo attenuation during double-talk in the ITU G.168 test 3B. The principle of robustness works at all stages of convergence of the robust algorithms. They resist divergence during double-talk, even in situations when they have not yet fully converged. It should also be mentioned that the performance loss due to the use of the nonlinearity for robustness is only minor. The robust PAPA algorithm reconverges faster than non-robust PNLMS for different tested speech sequences.

We have also described a new algorithm, SR-PNLMS, that simplifies the gradient calculations of the PNLMS algorithm by replacing the regressor vector with its sign. With proper normalization and regularization, this new algorithm has about the same convergence performance as the original PNLMS algorithm. However, it has lower complexity and smaller numerical range in its calculations. The latter feature is advantageous in a fixed-point implementation of the algorithm since higher precision can be maintained with a finite numerical range, or alternatively, fewer bits can be used without degrading performance.

The SR-PNLMS algorithm can be made robust to double-talk in the same way as previously shown for NLMS and PNLMS. The robust SR-PNLMS algorithm has been proven competitive to PNLMS and outperforms NLMS.

3. A Robust Fast Recursive Least-Squares Adaptive Algorithm

3.1 Introduction

Very often in the context of system identification, the error signal (e), which is by definition the difference between the system and model filter outputs, is assumed to be zero-mean, white, and Gaussian. In this case, the least-squares estimator is equivalent to the maximum likelihood estimator and, hence, it is asymptotically efficient. While this supposition is very convenient and extremely useful in practice, adaptive algorithms optimized on this basis may be very sensitive to minor deviations from the assumptions.

One good example of system identification with the above assumptions is network echo cancellation (EC) combined with a double-talk detector (DTD). Sometimes, the DTD fails to detect the beginning or end of a double-talk event and as a result, a burst of speech at the output of the echo path disturbs the estimation process (see Fig. 2.2). The occurrence rate of these bursts depends on the efficiency of the DTD and the intensity of double-talk. A desirable property of an adaptive algorithm is fast tracking. A high false alarm rate of the DTD reduces the amount of information that enters the algorithm, and therefore reduces the tracking rate. The false alarms should therefore be controlled so that valuable data are not discarded. Fewer false alarms, however, result in more detection misses and degradation of the transfer function estimate. To maintain tracking ability and high quality of the estimate, robustness against detection errors must be incorporated in the estimation algorithm itself. By *robustness* we mean insensitivity to small deviations of the the real distribution from the assumed model distribution [68].

Thus, the performance of an algorithm optimized for Gaussian noise could be very poor because of the unexpected number of large noise values that are not modeled by the Gaussian law. In our EC example, the probability density function (PDF) of the noise model should be a long-tailed PDF in order to take the bursts (due the DTD failure) into account [68], [72]. Therefore, we are interested in distributional robustness since the shape of the true underlying distribution can deviate from the assumed model (usually the Gaussian law).

As explained in [68], a robust procedure should have the following attributes:

- Reasonably good efficiency for the assumed model.
- Robust in the sense that small deviations from the model assumptions should only slightly impair the performance.
- Somewhat larger deviations from the model do not cause catastrophic failure.

In this study, we propose to use the following PDF:

$$
\begin{aligned}
p(z) &= \frac{1}{2} \exp\left\{-\ln\left[\cosh(\pi z/2)\right]\right\} \\
&= \frac{1}{2\cosh(\pi z/2)} ,
\end{aligned}
\tag{3.1}
$$

where the mean and the variance are respectively equal to 0 and 1. Figure 3.1 compares $p(z)$ to the Gaussian density:

$$
p_G(z) = \frac{1}{\sqrt{2\pi}} \exp\left\{-z^2/2\right\} .
\tag{3.2}
$$

We can see that $p(z)$ has a heavier tail than $p_G(z)$. Although, there does not appear to be a large deviation from the Gaussian PDF, if we take the

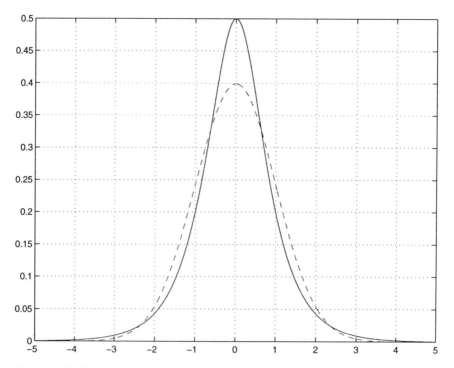

Fig. 3.1. PDFs of Gaussian density $p_G(z)$ (– –) and long-tailed density $p(z)$ (—), both with zero mean and unit variance

derivative of $\ln[p(z)]$ and $\ln[p_G(z)]$, it can easily be checked that $p'(z)$ is bounded while $p'_G(z)$ is not; this makes all the difference between a robust approach and a non-robust approach. Moreover, $p(z)$ has a high kurtosis and it is well-known that PDFs with large kurtosis are good models for speech signals, which is desired in our EC example.

In the following, we show how to derive a robust fast recursive least-squares adaptive algorithm[1] from (3.1) and how to apply it successfully to the problem of network echo cancellation. We show how to apply these ideas to the problem of acoustic echo cancellation in Chap. 6.

3.2 A Robust Fast Recursive Least-Squares Adaptive Algorithm

In the context of system identification, the error signal at time sample n between the system and model filter outputs is given by

$$e(n) = y(n) - \hat{y}(n) , \qquad (3.3)$$

where

$$\hat{y}(n) = \hat{\mathbf{h}}^T \mathbf{x}(n) \qquad (3.4)$$

is an estimate of the output signal $y(n)$,

$$\hat{\mathbf{h}} = \begin{bmatrix} \hat{h}_0 & \hat{h}_1 & \cdots & \hat{h}_{L-1} \end{bmatrix}^T$$

is the model filter, and

$$\mathbf{x}(n) = \begin{bmatrix} x(n) & x(n-1) & \cdots & x(n-L+1) \end{bmatrix}^T$$

is a vector containing the last L samples of the input signal x. Superscript ^{T} denotes transpose of a vector or a matrix.

Consider the following function:

$$J\left(\hat{\mathbf{h}}\right) = \rho \left[\frac{e(n)}{s(n)} \right] , \qquad (3.5)$$

where

$$\rho(z) = \ln[\cosh(z)]$$
$$= -\ln[2p(2z/\pi)] \qquad (3.6)$$

is a convex function and $s(n)$ is a positive scale factor more thoroughly described below. The gradient of $J\left(\hat{\mathbf{h}}\right)$ is:

---

[1] Least squares is a misnomer for this adaptive algorithm because we do not minimize the least-squares criterion but rather maximize a likelihood function. However, because of its familiarity, we continue to use this term here.

$$\nabla J\left(\hat{\mathbf{h}}\right) = -\frac{\mathbf{x}(n)}{s(n)}\psi\left[\frac{e(n)}{s(n)}\right] , \qquad (3.7)$$

where

$$\psi(z) = \frac{d\rho(z)}{dz}$$
$$= \tanh(z) . \qquad (3.8)$$

The second derivative (Hessian) of $J\left(\hat{\mathbf{h}}\right)$ is:

$$\nabla^2 J\left(\hat{\mathbf{h}}\right) = \frac{\mathbf{x}(n)\mathbf{x}^T(n)}{s^2(n)}\psi'\left[\frac{e(n)}{s(n)}\right] , \qquad (3.9)$$

where

$$\psi'(z) = \frac{1}{\cosh^2(z)} > 0 , \ \forall z . \qquad (3.10)$$

Figure 3.2 shows the functions $\rho(z)$, $\psi(z)$, and $\psi'(z)$.

Robust Newton-type algorithms have the following form (see [75] for the Newton algorithm):

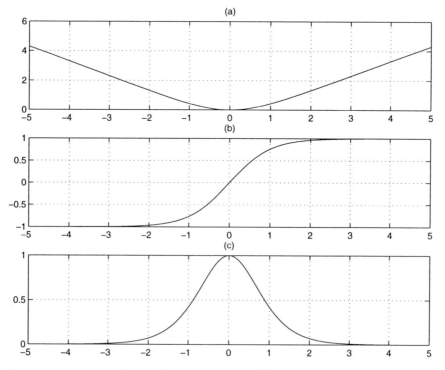

Fig. 3.2. Functions (a) $\rho(z)$, (b) $\psi(z)$, and (c) $\psi'(z)$

$$\hat{\mathbf{h}}(n) = \hat{\mathbf{h}}(n-1) - \mathbf{R}_{\psi'}^{-1}\nabla J\left[\hat{\mathbf{h}}(n-1)\right] , \tag{3.11}$$

where $\mathbf{R}_{\psi'}$ is an approximation of $E\left\{\nabla^2 J\left[\hat{\mathbf{h}}(n-1)\right]\right\}$ and $E\{\cdot\}$ denotes mathematical expectation. In this study we choose

$$\mathbf{R}_{\psi'} = \frac{\mathbf{R}}{s^2(n)}\psi'\left[\frac{e(n)}{s(n)}\right] , \tag{3.12}$$

with $\mathbf{R} = E\{\mathbf{x}(n)\mathbf{x}^T(n)\}$. This choice will allow us to derive a fast version of the algorithm. In practice, $\mathbf{R}$ is not known so we have to estimate it recursively:

$$\begin{aligned}
\mathbf{R}(n) &= \sum_{m=1}^{n} \lambda^{n-m}\mathbf{x}(m)\mathbf{x}^T(m) \\
&= \lambda\mathbf{R}(n-1) + \mathbf{x}(n)\mathbf{x}^T(n) ,
\end{aligned} \tag{3.13}$$

where λ $(0 < \lambda < 1)$ is an exponential forgetting factor.

We deduce a robust recursive least-squares (RLS) adaptive algorithm:

$$e(n) = y(n) - \hat{\mathbf{h}}^T(n-1)\mathbf{x}(n) , \tag{3.14}$$

$$\hat{\mathbf{h}}(n) = \hat{\mathbf{h}}(n-1) + \frac{s(n)}{\psi'\left[\frac{e(n)}{s(n)}\right]}\mathbf{k}(n)\psi\left[\frac{e(n)}{s(n)}\right] , \tag{3.15}$$

where

$$\mathbf{k}(n) = \mathbf{R}^{-1}(n)\mathbf{x}(n) \tag{3.16}$$

is the Kalman gain [66]. From Fig. 3.2, we can see that $0 < \psi'(z) \le 1$, $\forall z$, and $\psi'(z)$ can become very small. In order to avoid divergence of the robust RLS, we do not allow $\psi'(z)$ to be lower than 0.5. So, in practice, we compute $\psi'(z)$ according to (3.10) but limit it to 0.5 if it is lower than 0.5. From (3.15) it can be understood that large errors will be limited by the function $\psi(\cdot)$. Note that if we choose $\rho(z) = z^2$, then $\psi(z) = 2z$ and $\psi'(z) = 2$, and the algorithm is exactly the non-robust RLS [66].

A robust fast RLS (FRLS) can be derived by using the *a priori* Kalman gain $\mathbf{k}'(n) = \mathbf{R}^{-1}(n-1)\mathbf{x}(n)$. This *a priori* Kalman gain can be computed recursively with $5L$ multiplications, and the error, as well as the adaptation parts, in $2L$ multiplications. "Stabilized" versions of FRLS (with L more multiplications) exist in the literature but they are not much more stable than their non-stabilized counterparts with non-stationary signals like speech. Our approach to fix this problem is simply to re-initialize the predictor-based variables when instability is detected using the maximum likelihood variable, which is an inherent variable of the FRLS. This method has worked very well in all of the simulations that have been done.

One other important part of the algorithm is the estimate of the scale factor s. Traditionally, the scale factor is used to make a robust algorithm invariant to the noise level. It should reflect the minimum mean-square error, be robust to short burst disturbances (double-talk in our application), and track longer changes of the residual error (echo path changes). We have chosen the scale factor estimate as

$$s(n+1) = \lambda_s s(n) + (1 - \lambda_s) \frac{s(n)}{\psi' \left[\frac{e(n)}{s(n)} \right]} \left| \psi \left[\frac{e(n)}{s(n)} \right] \right| , \qquad (3.17)$$

$$s(0) = \sigma_x ,$$

which is very simple to implement. This method of estimating s is justified in [51] (and discussed in Chap. 2). With this choice, the current estimate of s is governed by the level of the error signal in the immediate past over a time interval roughly equal to $1/(1 - \lambda_s)$. When the algorithm has not yet converged, s is large. Hence the limiter is in its linear portion and therefore the robust algorithm behaves roughly like the conventional RLS algorithm. When double-talk occurs, the error is determined by the limiter and by the scale of the error signal during the recent past in the absence of double-talk. Thus, the divergence rate is reduced for double-talk bursts of length less than about $1/(1 - \lambda_s)$. This gives ample time for the DTD to act. If there is a system change, the algorithm will not track immediately. However, as the scale estimator tracks the larger error signal, the nonlinearity is scaled up and the convergence rate accelerates. The trade-off between robustness and tracking rate of the adaptive algorithm is thus governed by the tracking rate of the scale estimator, which is controlled by a single parameter λ_s.

In Table 3.1, we give a robust FRLS algorithm with a complexity of $O(7L)$.

3.3 Application to Network Echo Cancellation and Simulations

In telephone networks that involve connection of 4-wire and 2-wire links, an echo is generated at the hybrid. This echo has a disturbing influence on the conversation and must therefore be canceled. Figure 3.3 shows the principle of a network echo canceler (EC). The far-end speech signal $x(n)$ goes through the echo path represented by a filter $\mathbf{h}$, then it is added to the near-end talker signal $v(n)$ and ambient noise $w(n)$. The composite signal is denoted $y(n)$. Most often the echo path is modeled by an adaptive FIR filter, $\hat{\mathbf{h}}(n)$, which subtracts a replica of the echo and thereby achieves cancellation. This may look like a simple straightforward system identification task for the adaptive filter. However, in most conversations there are so-called *double-talk* situations that make the identification much more problematic than might appear at first glance. Double-talk occurs when the two talkers on both sides

Table 3.1 A robust FRLS algorithm

Prediction:

$$e_a(n) = x(n) - \mathbf{a}^T(n-1)\mathbf{x}(n-1)$$

$$\varphi_1(n) = \varphi(n-1) + e_a^2(n)/E_a(n-1)$$

$$\begin{bmatrix} \mathbf{t}(n) \\ m(n) \end{bmatrix} = \begin{bmatrix} 0 \\ \mathbf{k}'(n-1) \end{bmatrix} + \begin{bmatrix} 1 \\ -\mathbf{a}(n-1) \end{bmatrix} e_a(n)/E_a(n-1)$$

$$E_a(n) = \lambda[E_a(n-1) + e_a^2(n)/\varphi(n-1)]$$

$$\mathbf{a}(n) = \mathbf{a}(n-1) + \mathbf{k}'(n-1)e_a(n)/\varphi(n-1)$$

$$e_b(n) = E_b(n-1)m(n)$$

$$\mathbf{k}'(n) = \mathbf{t}(n) + \mathbf{b}(n-1)m(n)$$

$$\varphi(n) = \varphi_1(n) - e_b(n)m(n)$$

$$E_b(n) = \lambda[E_b(n-1) + e_b^2(n)/\varphi(n)]$$

$$\mathbf{b}(n) = \mathbf{b}(n-1) + \mathbf{k}'(n)e_b(n)/\varphi(n)$$

Filtering:

$$e(n) = y(n) - \hat{\mathbf{h}}^T(n-1)\mathbf{x}(n)$$

$$\psi\left[\frac{e(n)}{s(n)}\right] = \tanh\left[\frac{e(n)}{s(n)}\right]$$

$$\psi'\left[\frac{e(n)}{s(n)}\right] = 1/\cosh^2\left[\frac{e(n)}{s(n)}\right]$$

$$\psi'_f\left[\frac{e(n)}{s(n)}\right] = \begin{cases} \psi'\left[\frac{e(n)}{s(n)}\right] & \text{if } \psi'\left[\frac{e(n)}{s(n)}\right] \geq 0.5 \\ 0.5 & \text{otherwise} \end{cases}$$

$$\hat{\mathbf{h}}(n) = \hat{\mathbf{h}}(n-1) + \frac{s(n)}{\psi'_f\left[\frac{e(n)}{s(n)}\right]\varphi(n)}\mathbf{k}'(n)\psi\left[\frac{e(n)}{s(n)}\right]$$

$$s(n+1) = \lambda_s s(n) + (1-\lambda_s)\frac{s(n)}{\psi'_f\left[\frac{e(n)}{s(n)}\right]}\left|\psi\left[\frac{e(n)}{s(n)}\right]\right|$$

speak simultaneously, i.e. $x(n) \neq 0$ and $v(n) \neq 0$. In this situation, the near-end speech acts as a high-level uncorrelated noise to the adaptive algorithm. The disturbing near-end speech may therefore cause the adaptive filter to diverge, passing annoying audible echo through to the far end. A common way to alleviate this problem is to slow down or completely halt the filter adaptation when near-end speech is detected. This is the very important role of the DTD.

In this section, we wish to compare, by way of simulation, the robust and non-robust FRLS algorithms in the context of network EC with a DTD. In these simulations, we use the Geigel DTD (see Chap. 2) [37] in which the

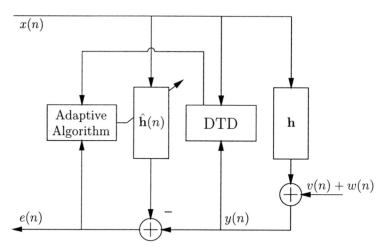

Fig. 3.3. Block diagram of the echo canceler and double-talk detector

settings are chosen the same as commonly used in commercial hardware and assumes a 6 dB detection threshold. The hybrid attenuation is 20 dB and the length of the echo path $\mathbf{h}$ is $L = 512$. The same length is used for the adaptive filter $\hat{\mathbf{h}}(n)$. The sampling rate is 8 kHz and the echo-to-ambient-noise ratio is equal to 39 dB. We have chosen $\lambda_s = 0.992$ for the scale estimate $s(n)$, and $s(n)$ is never allowed to be lower than 0.01. For the adaptive algorithms, we used $\lambda = 1 - 1/(3L)$.

An example of the performance of the robust and non-robust FRLS algorithms during double-talk when speech is used is shown in Fig. 3.4. The far-end speaker is female, the near-end speaker is male, and the average far-end to near-end ratio is 6 dB. The divergence rate of the algorithms does not strongly depend on the power of the near-end signal. We can see that even when the DTD is used, the non-robust FRLS diverges because the DTD does not react fast enough, while for the robust FRLS there is only a slight increase of the misalignment (difference between the true echo path and the estimated path).

Figure 3.5 shows the behavior after an abrupt system change where the impulse response is shifted 200 samples at 1.5 seconds. The re-convergence rate of the robust algorithm is a little bit slower than the non-robust version, but this is a small price to pay for robustness against double-talk. Note that since the FRLS algorithm converges more rapidly than other algorithms like NLMS, this somewhat slower convergence is still fully acceptable.

3.4 Conclusions

In robust statistics, the function $\psi(\cdot)$ which can be directly derived from the model distribution of the error signal plays a key role. If we want to

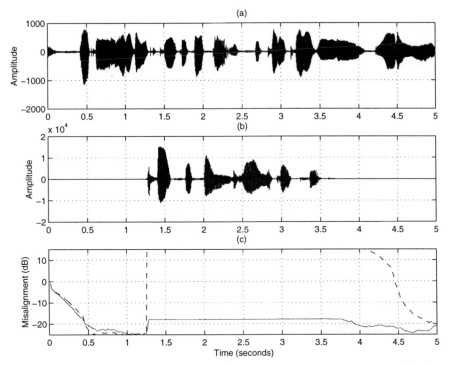

Fig. 3.4. Double-talk situation. (a) Far-end signal. (b) Near-end signal. (c) Misalignment of the robust FRLS (—) and non-robust FRLS (– –) algorithms

make a robust estimate that has good efficiency, we should choose a ψ that is bounded. In this chapter, we proposed to use $\psi(z) = \tanh(z)$ but other choices are possible such as the Huber function [68]. We have shown how to derive robust Newton-type algorithms and from that we have derived a robust RLS algorithm and its fast version. We have also shown that the robust FRLS algorithm behaves very nicely when it is used for network echo cancellation, even with a DTD (e.g., Geigel algorithm) that fails to detect a double-talk situation quickly, which is almost always the case in practice.

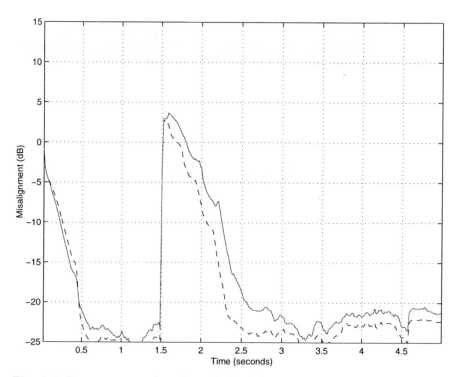

Fig. 3.5. Reconvergence after abrupt hybrid change. Misalignment of the robust FRLS (—) and non-robust FRLS (– –) algorithms

4. Dynamic Resource Allocation for Network Echo Cancellation

4.1 Introduction

Current adaptation algorithms for network echo cancelers are designed without regard to the fact that, invariably, a single canceler chip handles many conversations simultaneously. This implies that for N_c channels, the processor must handle N_c times the *peak* computational load of a single channel. If the number of channels is large, however, it should be possible to reduce the demands on the processor to something close to N_c times the *average* load. Some additional computational capacity would, of course, be necessary to take care of statistical fluctuation in the requirements, but the required safety margin becomes smaller as N_c becomes larger. (With the speed and memory now available on a chip, the number of channels can be several hundred, so the safety margin might not have to be large.) Once the problem is looked upon as that of dealing with a large number of channels, it is also possible to take advantage of other knowledge about speech patterns and characteristics of long distance circuits to further reduce the computational load. In this chapter, we show how the computational requirement can, in principle, be reduced by a very large factor — perhaps as large as thirty.

Basically we capitalize on three facts. First, during a telephone conversation, there are many pauses in each speech signal. These pauses have been exploited to decrease the idle time in telephone connections, so called TASI (time assignment speech interpolation) networks, since the 1960s [25]. An echo canceler can also take advantage of these pauses. Second, network echo paths are sparse, i.e. only a small number of coefficients are nonzero. By utilizing this sparseness property of the responses, it is possible to increase the convergence rate and decrease the complexity of adaptive filters [73], [86], [124], [67]. Finally, echo paths do not change much during a conversation. Hence the adaptive filter need not be updated continuously. It is estimated that it needs to be updated perhaps only 10% of the time. These three features can be exploited to design an efficient multiple-channel echo cancellation system.

The rest of the chapter is organized as follows: In Sect. 4.2, the PNLMS algorithm is presented. Section 4.3 gives some proposals on how to simplify the PNLMS algorithm, and shows the reduction in complexity that may be achieved. Performance simulations are presented in Sect. 4.4, and Sect. 4.5 gives a discussion of the results and problems associated with this approach.

4.2 The PNLMS and PNLMS++ Algorithms

In this section, we only give the algorithms; for more details, the reader is referred to Chap. 2. Figure 3.3 (Chap. 3) shows the various signals of interest. In derivations and descriptions, the following notation is used for these quantities:

$$x(n) = \text{ Far-end signal,}$$
$$y(n) = \text{ Echo and background noise,}$$
$$\text{possibly including near-end signal,}$$
$$\mathbf{x}(n) = [x(n) \;\cdots\; x(n-L+1)]^T \text{ , Excitation vector,}$$
$$\mathbf{h} = [h_0 \;\cdots\; h_{L-1}]^T \text{ , True echo path,}$$
$$\hat{\mathbf{h}}(n) = [\hat{h}_0(n) \;\cdots\; \hat{h}_{L-1}(n)]^T \text{ , Estimated echo path.}$$

Here L is the length of the adaptive filter, and n is the time index. In the PNLMS algorithm [38], an adaptive individual step-size is assigned to each filter coefficient. The step-sizes are calculated from the last estimate of the filter coefficients in such a way that a larger coefficient receives a larger increment, thus increasing the convergence rate of that coefficient. This has the effect that active coefficients are adjusted faster than non-active coefficients (i.e., small or zero coefficients). Hence, PNLMS converges much faster than NLMS for sparse impulse responses (i.e., responses in which only a small percentage of coefficients is significant). Most impulse responses in the telephone network have this characteristic.

The PNLMS algorithm is described by the following equations:

$$e(n) = y(n) - \hat{\mathbf{h}}^T(n-1)\mathbf{x}(n) , \tag{4.1}$$

$$\hat{\mathbf{h}}(n) = \hat{\mathbf{h}}(n-1) + \frac{\mu}{\mathbf{x}^T(n)\mathbf{G}(n)\mathbf{x}(n) + \delta}\mathbf{G}(n)\mathbf{x}(n)e(n) , \tag{4.2}$$

$$\mathbf{G}(n) = \text{diag}\{g_0(n),\ldots,g_{L-1}(n)\} . \tag{4.3}$$

Here $\mathbf{G}(n)$ is a diagonal matrix that adjusts the step-sizes of the individual taps of the filter, μ is the overall step-size parameter, and δ is a regularization parameter which prevents division by zero and stabilizes the solution when speech is used as the input (far-end) signal. The diagonal elements of $\mathbf{G}(n)$ are calculated as follows [38]:

$$\gamma_l(n) = \max\{\rho\max[\delta_{\mathrm{p}}, |\hat{h}_0(n-1)|, \dots, |\hat{h}_{L-1}(n-1)|], |\hat{h}_l(n-1)|\}\,,$$
(4.4)

$$g_l(n) = \frac{\gamma_l(n)}{\dfrac{1}{L}\displaystyle\sum_{i=0}^{L-1}\gamma_i(n)}\,, \quad 0 \le l \le L-1\,.$$
(4.5)

Parameters δ_{p} and ρ are positive numbers with typical values $\delta_{\mathrm{p}} = 0.01$, $\rho = 5/L$. See Chap. 2 for further explanation and discussion.

A variant of PNLMS is the PNLMS++ algorithm [54]. In this algorithm, for odd-numbered time steps the matrix $\mathbf{G}(n)$ is calculated as above, while for even-numbered steps, $\mathbf{G}(n)$ is the identity matrix

$$\mathbf{G}(n) = \mathbf{I}\,,$$
(4.6)

which results in an NLMS iteration. Alternating between NLMS and PNLMS iterations has several advantages over using PNLMS alone. For example, it makes the PNLMS++ algorithm less sensitive to the assumption of a sparse impulse response, without unduly sacrificing performance.

4.3 DRA Algorithm

In this section, we outline the principles of a simplified algorithm that exploits the sparseness of network echo paths and the fact that the adaptive filter does not need to be updated when the residual error is small. We call this the *dynamic resource allocation* (DRA) algorithm. We also present the theoretical complexity gains one can achieve. Any overhead that depends on the choice of implementation platform is not considered in this chapter. It is assumed that one or more computation engines serve the channels and some logic has been designed to control and distribute the resources. Decisions, e.g., which channels should be updated, are based on results from voice activity and double-talk detection.

The PNLMS algorithm takes advantage of the sparseness of the impulse response to improve the convergence rate, but its complexity is greater by a factor of 2 compared to that of the NLMS algorithm. In the following we show that we can do much better than that.

4.3.1 A Simple Algorithm to Update only Active Channels and Coefficients

In a two-way conversation, each talker is active only about half of the time; additionally, there are pauses between sentences and syllables. During these inactive time slots, no coefficient updating is needed. Furthermore, since network echo path responses are sparse, we can focus computations on only the active (non-zero) coefficients. The following algorithm, which may be seen as a simplified PNLMS++ algorithm, saves a large number of multiplications at the expense of some additional overhead. The key differences compared to the current implementations of PNLMS++ are:

1. No coefficient is updated if the far-end signal is inactive or double-talk has been detected.
2. No coefficient is updated if the residual error is sufficiently small.
3. Step-sizes for the active taps are made equal instead of the step-sizes specified by the matrix $\mathbf{G}(n)$. This is equivalent to quantizing $\mathbf{G}(n)$ to a one-bit representation, whereby $g_l(n)$ is either of value $1/L_a$ (active) or 0 (inactive). (See below for the definition of L_a.)
4. All coefficients are updated (i.e., an NLMS iteration is made) every Mth iteration. Only the active coefficients are updated at all other iterations.
5. The index of active coefficients is updated every Mth iteration.

Periods of inactivity are easily identified with a look-ahead of one, or a few, samples at the outputs of the voice activity and double-talk detectors. Hence the first item is easily implemented. The implementation of item 2 is explained in the next subsection. Items 3–5 are implemented as follows:

Let us first define an "active set," i.e., the set of active tap weights. To this end, define a threshold T_a, and sort the tap weights in descending order of absolute value. Then define the active set A as the first L_a weights in this list, such that their cumulative magnitude just exceeds T_a times the cumulative magnitude of *all* the L taps of the filter. In symbols,

$$L_a = \min_k \left\{ k : T_a \sum_{l=1}^{L} |\hat{h}_{(l)}(n)| \leq \sum_{l=1}^{k} |\hat{h}_{(l)}(n)| \right\} , \tag{4.7}$$

$$A = \{l : l \leq L_a\} , \tag{4.8}$$

where $\hat{h}_{(l)}$ denotes the coefficient with the lth largest magnitude. The threshold T_a is selected in the range $T_a = [0.9, 1[$. With $T_a = 1$, the active set covers all the taps and we get the standard NLMS algorithm. From an implementation point of view, it may be appropriate to limit the maximum size of the set to L_{max} thus $L_a \leq L_{max} < L$. The maximum load of a channel can thereby be limited. For $M - 1$ consecutive iterations, the coefficients of the active set are updated as follows:

$$e(n) = y(n) - \sum_{l \in A} \hat{h}_l(n-1)x(n-l) , \tag{4.9}$$

$$\hat{h}_l(n) = \hat{h}_l(n-1) + \frac{\mu}{\sum_{l \in A} x^2(n-l) + \delta} x(n-l)e(n) , \; l \in A . \tag{4.10}$$

Every Mth iteration a full NLMS iteration is made, i.e.,

$$e(n) = y(n) - \hat{\mathbf{h}}^T(n-1)\mathbf{x}(n) , \tag{4.11}$$

$$\hat{\mathbf{h}}(n) = \hat{\mathbf{h}}(n-1) + \frac{\mu}{\mathbf{x}^T(n)\mathbf{x}(n) + \delta} \mathbf{x}(n)e(n) . \tag{4.12}$$

Increasing M will reduce the average complexity but also worsen the tracking performance. In the simulations described in Sect. 4.4, we used $M = 10$.

4.3.2 Stopping Adaptation when Small Residual Error is Detected

Echo paths on the network vary slowly, in general. Hence adaptation is needed only at a small percentage of iterations, perhaps no more than 10%. This would yield a large reduction in computation, since on a vast majority of iterations we need to compute only the convolution with the small number (L_a) of coefficients in the set A. Therefore, we propose that when the error signal is sufficiently small, we do not update or sort the tap weights. Asymptotically, for network echo cancelers, the complexity of this algorithm would thus be reduced essentially to the computation of a convolution on the active taps only.

A good decision variable, to decide if the residual error is small enough, is the normalized mean-square error defined as follows:

$$\xi_r(n) = \frac{< e^2(n) >_{N_r}}{< y^2(n) >_{N_r} + \delta_y},$$ (4.13)

where

$$< e^2(n) >_{N_r} = \frac{1}{N_r} \sum_{n-N_r+1}^{n} e^2(n)$$ (4.14)

is the mean-square error,

$$< y^2(n) >_{N_r} = \frac{1}{N_r} \sum_{n-N_r+1}^{n} y^2(n)$$ (4.15)

is the mean-square output, δ_y is a regularization parameter which prevents division by zero during silences between words, and N_r is the length of the window used to estimate energy. The window length N_r should preferably be chosen no larger than necessary, in order not to degrade the tracking performance of the adaptive algorithm when the echo path changes.

The adaptive algorithm proceeds as follows: at each iteration n, (4.13) is computed and $\xi_r(n)$ is compared to a threshold T_r (a typical range is -40 to $-30\,\text{dB}$). The decision rule is simple: if $\xi_r(n) \geq T_r$, then the residual error is not considered small enough and the algorithm continues to update; If $\xi_r(n) < T_r$, then the residual error is considered negligible and the algorithm neither updates nor sorts the tap weights of the filter. However, in all cases, the convolution on the active taps is still performed.

4.3.3 Theoretical Reduction in Complexity

The complexity of the DRA algorithm is compared to a PNLMS implementation with respect to multiplications and other required computations. Computations required for the various steps of the proposed algorithm are:

Eq. (4.8): $k_0 L \log_2(L) + 2L + L_a$ (sorting instructions) [1],
Eq. (4.9): L_a (multiplications),
Eq. (4.10): $2L_a$ (multiplications),
Eq. (4.11): L (multiplications),
Eq. (4.12): $2L$ (multiplications).

Let the probability of active speech be denoted by p_s and the probability of active adaptation by p_a. Then, assuming equal weight for multiplications and sorting instructions, the average required number of computations is

$$C_0 = p_s p_a \frac{3L}{M} \ , \text{ full NLMS update,} \tag{4.16}$$

$$C_1 = p_s \frac{(M-1)(1+2p_a)L_a}{M} \ , \text{ update of active coefficients,} \tag{4.17}$$

$$C_2 = p_s p_a \frac{L + L_a + k_0 L \log_2(L)}{M} \ , \text{ update of the active set,} \tag{4.18}$$

and the average total number of computations is

$$C_{\text{DRA}} = C_0 + C_1 + C_2 \ . \tag{4.19}$$

For comparison, note that the implementation of PNLMS++ via equations (4.1-4.5) requires:

Eq. (4.1): L (multiplications),
Eq. (4.2): $3L$ (multiplications),
Eq. (4.5): $2L$ (comparisons).

Again, with equal weight for multiplications and comparisons, the total number of computations[2] is

$$C_{\text{PNLMS++}} = 6L \ . \tag{4.20}$$

As an illustration, assuming the expected typical values

$$p_s \approx 0.5 \ ,$$
$$p_a \approx 0.1 \ ,$$
$$L_a = 100 \ ,$$
$$k_0 \approx 1 \ ,$$
$$M = 10 \ ,$$
$$L = 768 \ ,$$

we find that

$$C_{\text{DRA}} \approx 70 \ ,$$
$$C_{\text{PNLMS++}} = 4608 \ ,$$

---

[1] This assumes that an algorithm like *quicksort* is used, where k_0 is a proportionality constant for the sorting algorithm.
[2] Strictly speaking, this is the complexity of PNLMS. For PNLMS++ some of the computations can be eliminated at all even-numbered iterations.

which shows that the average complexity of the proposed algorithm could be drastically less than that of PNLMS. However, for various reasons, the estimate in (4.21) should not be taken literally. First, it does not allow for a safety margin to accommodate occasional statistical outliers, and it is not yet clear how large that needs to be made. Second, at present there is no hard evidence to justify the estimate $p_a \approx 0.1$. Third, the estimate of complexity of the sorting algorithm is not rigorous. Finally, the choice of $M = 10$, which controls the initial convergence rate, has not yet been optimized. Nevertheless, a reduction in complexity by a factor as large as 25 or 30 appears to be possible.

4.3.4 Overload Management

As mentioned earlier, a safety margin must be provided to allow for statistical fluctuations in the computational load. However, even with a safety margin, there is a finite probability of temporarily running out of resources. Some possibilities to handle such an overload situation are:

- Not perform coefficient updates on some of the active channels (or increase the level of the threshold T_r);
- Neither perform coefficient updates nor calculate residual echoes of some of the active channels. Just apply echo suppression.

4.4 Simulations

International Telecommunication Union Recommendation ITU-T G.168 [69] specifies certain test signals to be used for evaluating the performance of echo cancelers. The signal used is the so called composite source signal (CSS), which has properties similar to those of speech with both voiced and unvoiced sequences as well as pauses. In the first part of our simulations, we use the CCS as an excitation signal and show the convergence of the DRA algorithm without halting adaptation (due to inactive far-end signal, double-talk detection, or small residual error). In these simulations, results from three kinds of echo paths are shown: a sparse, a dispersive, and a multireflection path. These three paths represent a variety of responses that can be expected in practice. Figure 4.1 shows these impulse responses and their frequency response magnitude functions. The far-end signal (CSS) used is shown in Fig. 4.2(a). The parameter settings chosen for the following simulations are:

- $\mu = 0.2$, $L = 512$ (64 ms) , $\delta = 2 \cdot 10^5$, $\delta_p = 0.01$, $\rho = 0.001$.
- $T_a = 0.98$, $L_{max} = 200$.
- $\sigma_x = 2100$, ENR ≈ 37 dB (echo-to-PCM quantization noise ratio).
- Hybrid attenuation: 6, 11 dB .
- $\mathbf{h}(-1) = \mathbf{0}$.

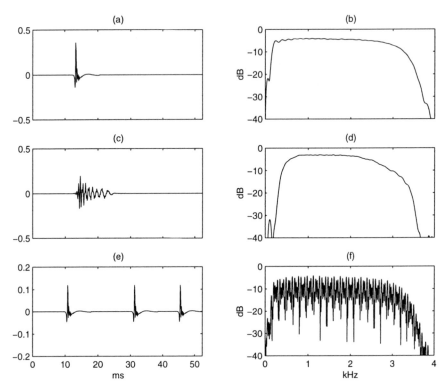

Fig. 4.1. Description of three hybrids used in the simulations. (a), (c), (e) Impulse responses. (b), (d), (f) Corresponding frequency response magnitudes

The performance is evaluated by using the mean-square error (MSE) which is given by,

$$\text{MSE} = \frac{\text{LPF}\{e^2(n)\}}{\text{LPF}\{y^2(n)\}} \, . \tag{4.21}$$

Here LPF denotes a lowpass filter, chosen here to be a leaky integrator with a single pole at 0.999 (time constant of $1/8$ s). This choice does not significantly affect the convergence curves shown in the figures. Figures 4.2(b)–(d) show the convergence behavior of the DRA algorithm compared to PNLMS++ and NLMS. The DRA algorithm has somewhat poorer initial convergence rate than PNLMS++ but is still considerably faster than NLMS. With longer echo paths, the performance improvement compared to NLMS will be even greater.

In the second part of the simulations, we use an actual speech signal as the excitation and show the convergence and tracking of the DRA algorithm when adaptation is halted due to sufficiently low residual error. (Halting due to inactive far-end signal or double-talk detection was not considered here.) The following parameters are used:

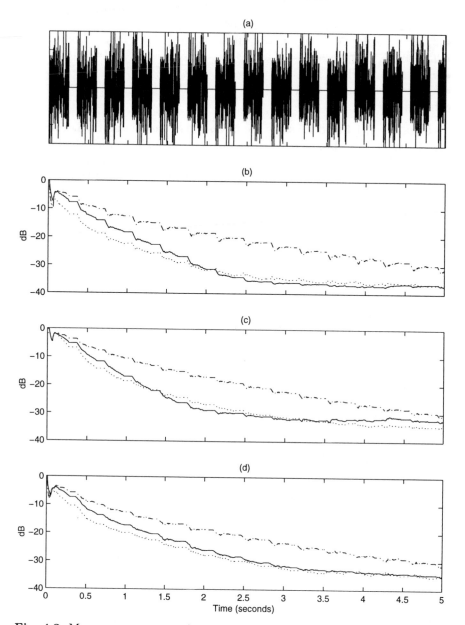

Fig. 4.2. Mean-square error performance for composite source signal. (a) Far-end CSS signal. (b) Sparse hybrid. (c) Disperse hybrid. (d) Hybrid with three reflections. Solid line - DRA algorithm; dashed line - NLMS; dotted line - PNLMS++

- $\mu = 0.2$, $L = 512$ (64 ms) , $\delta = 4 \cdot 10^6$.
- $T_r = -38.5\,\mathrm{dB}$, $N_r = 40$, $\delta_y = 1 \cdot 10^6$.
- $T_a = 0.98$, $L_{\max} = 200$.
- $\sigma_x = 1900$, ENR $\approx 39\,\mathrm{dB}$.
- Hybrid attenuation: 6, 11 dB .
- $\mathbf{h}(-1) = \mathbf{0}$.

Figures 4.3, 4.4, and 4.5 show the mean-squared error and the normalized misalignment, $\|\mathbf{h} - \hat{\mathbf{h}}\|/\|\mathbf{h}\|$, for three different algorithms when the input is speech and the echo path is the one shown in Fig. 4.1(a). In Fig. 4.3, the NLMS algorithm is used. In Fig. 4.4, we used the DRA algorithm without stopping adaptation during periods of small residual error (i.e. with T_r set to $-\infty$). In Fig. 4.5, the same algorithm was used but with T_r selected such that the adaptation was halted more that 55% of the time. We can see that the DRA algorithm (whether or not adaptation is stopped) outperforms the NLMS algorithm.

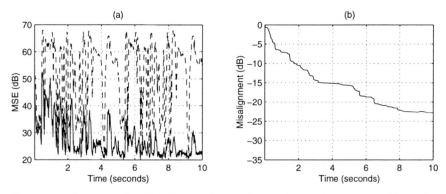

Fig. 4.3. Behavior of the NLMS algorithm with speech signal as input and hybrid of Fig. 4.1(a). (a) MSE (–) as compared to original echo level (– –). (b) Misalignment

It is also of interest to see what happens in a tracking situation, i.e., when the echo path changes. Figures 4.6, 4.7, and 4.8 show the results of repeating the experiments of Figs. 4.3–4.5, except that the echo path is that of Fig. 4.1(a) for the first 10 seconds and that of Fig. 4.1(c) for the next ten seconds. Again, we can see that the proposed algorithm (whether or not adaptation is stopped) tracks better than the NLMS algorithm.

4.5 Discussion

In this chapter, a dynamic resource allocation algorithm has been proposed for decreasing the complexity of the adaptive algorithm in a multiple-channel

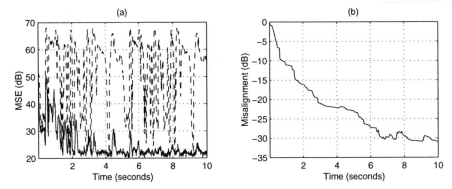

Fig. 4.4. Behavior of the DRA algorithm, without stopping adaptation, with speech signal as input and hybrid of Fig. 4.1(a). (a) MSE (–) as compared to original echo level (– –). (b) Misalignment

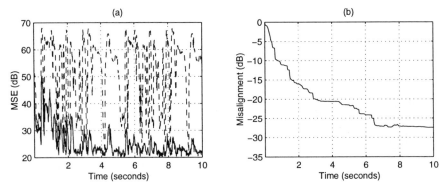

Fig. 4.5. Behavior of the DRA algorithm, where adaptation is stopped more than 55% of the time, with speech signal as input and hybrid of Fig. 4.1(a). (a) MSE (–) as compared to original echo level (– –). (b) Misalignment

echo canceler system. Emphasis was placed on finding simple procedures for choosing active regions of the impulse response and halting adaptation when the residual error is small. Inhibiting adaptation for far-end signal inactivity and/or double-talk detection also contributes to the computational efficiency of the DRA algorithm in a multiple-channel implementation. Though more careful analysis and development of the algorithm is needed, these proposals give some idea of what can be done from an algorithmic point of view in order to improve the efficiency of the implementation.

An important aspect of the problem that we have not discussed in the chapter is the possibility of reducing the requirement of storage capacity. Reduction in storage requirements is necessary if the reduction of computational complexity is to be fully exploited.

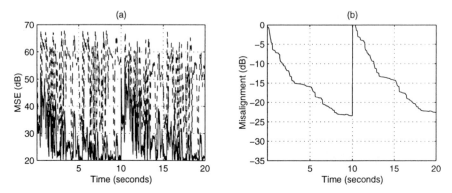

Fig. 4.6. Behavior of the NLMS algorithm with speech signal as input and when the echo path changes at time 10 s from the one in Fig. 4.1(a) to the one in Fig. 4.1(c). (a) MSE (–) as compared to original echo level (– –). (b) Misalignment

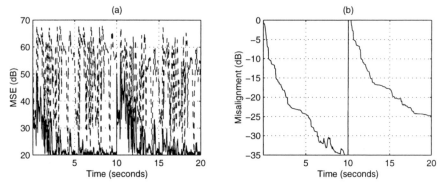

Fig. 4.7. Behavior of the DRA algorithm, without stopping adaptation, with speech signal as input and when the echo path changes at time 10 s from the one in Fig. 4.1(a) to the one in Fig. 4.1(c). (a) MSE (–) as compared to original echo level (– –). (b) Misalignment

One possibility for reducing storage requirements is to store the L_a coefficients in the set A with full precision, and the rest with reduced precision. Another possibility is to store L_max coefficients with full precision and the rest with reduced precision. Since the inactive coefficients are, in general, much larger in number, this procedure can significantly reduce the memory requirement. What remains to be seen is how few bits can be used for the inactive taps without degrading performance. Also, it needs to be determined whether this scheme can be implemented without a large increase in overhead.

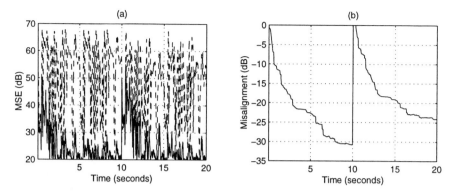

Fig. 4.8. Behavior of the DRA algorithm, where adaptation is stopped more than 45% of the time, with speech signal as input and when the echo path changes at time 10 s from the one in Fig. 4.1(a) to the one in Fig. 4.1(c). (a) MSE (–) as compared to original echo level (– –). (b) Misalignment

5. Multichannel Acoustic Echo Cancellation

5.1 Introduction

One may ask a legitimate question: why do we need multichannel sound for telecommunication? Let's take the following example. When we are in a room with several people talking, laughing, or just communicating with each other, thanks to our binaural auditory system, we can concentrate on one particular talker (if several persons are talking at the same time), localize or identify a person who is talking, and somehow we are able to process a noisy or a reverberant speech signal in order to make it intelligible. On the other hand, with only one ear or, equivalently, if we record what happens in the room with one microphone and listen to this monophonic signal, it will likely make all of the above mentioned tasks more difficult. So, multichannel sound teleconferencing systems provide a realistic presence that mono-channel systems cannot offer.

In such hands-free systems, multichannel acoustic echo cancelers (MCAECs) are absolutely necessary for full-duplex communication [121]. Let P be the number of channels. For a teleconferencing system, the MCAECs consist of P^2 adaptive filters aiming at identifying P^2 echo paths from P loudspeakers to P microphones. We assume that the teleconferencing system is organized between two rooms: the "transmission" and "receiving" rooms. The transmission room is sometimes referred to as the far-end and the receiving room as the near-end. So each room needs an MCAEC for each microphone. In the following, we detail the concept on one microphone in the receiving room, knowing that the same approach applies to the other microphones. Thus, multichannel acoustic echo cancellation consists of direct identification of a multi-input, unknown linear system, consisting of the parallel combination of P acoustic paths $(h_1, h_2, ..., h_P)$ extending through the receiving room from the loudspeakers to the microphone. The multichannel AEC tries to model this unknown system by P adaptive filters $(\hat{h}_1, \hat{h}_2, ..., \hat{h}_P)$.

Although conceptually very similar, multichannel acoustic echo cancellation (MCAEC) is fundamentally different from traditional mono echo cancellation in one respect: a straightforward generalization of the mono echo canceler would not only have to track changing echo paths in the receiving room, *but also in the transmission room*! For example, the canceler would have to reconverge if one talker stops talking and another starts talking at a different location in the transmission room. There is no adaptive algorithm that can track such a change sufficiently fast and this scheme therefore results in poor echo suppression. Thus, a generalization of the mono AEC in the multichannel case does not result in satisfactory performance.

The theory explaining the problem of MCAEC was described in [121] and [19]. The fundamental problem is that the multiple channels may carry linearly related signals which in turn may make the normal equation to be solved by the adaptive algorithm singular. This implies that there is no unique solution to the equation but an infinite number of solutions, and it can be shown that all but the true one depend on the impulse responses of the transmission room. As a result, intensive studies have been made of how to handle this properly. It is shown in [19] that the only solution to the nonuniqueness problem is to reduce the coherence between the different signals, and an efficient low complexity method for this purpose is also given.

Lately, attention has been focused on the investigation of other methods that decrease the cross-correlation between the channels in order to get well behaved estimates of the echo paths [50], [58], [112], [1], [71]. The main problem is how to reduce the coherence sufficiently without affecting the stereo perception and the sound quality.

The performance of the MCAEC is more severely affected by the choice of algorithm than the monophonic counterpart [10], [49]. This is easily recognized since the performance of most adaptive algorithms depends on the condition number of the input signal covariance matrix. In the multichannel case, the condition number is very high; as a result, algorithms such as the least mean square (LMS) or the normalized LMS (NLMS), which do not take into account the cross-correlation among all the input signals, converge very slowly to the true solution. It is therefore highly interesting to study multichannel adaptive filtering algorithms. Straightforward extensions of single-channel algorithms may not be the best choice for the MCAEC application. Standard algorithms (and their fast versions, i.e. low complexity versions) are the recursive least-squares (RLS), LMS, and affine projection algorithm (APA). A framework for multichannel adaptive filtering can be found in [8]. Results from this and related papers will also be discussed below.

This chapter is organized as follows: Sect. 5.2 describes the fundamental difference between mono and multichannel acoustic echo cancellation and explains the nonuniqueness problem. Section 5.3 discusses the impulse response tail effect. In Sect. 5.4, we show the link between the coherence function and the covariance matrix of the input signals. In Sect. 5.5, a summary of recent research on methods for decorrelating the stereo channels is given. Section 5.6 reviews conventional multichannel time-domain adaptive filter algorithms suitable for MCAEC. In Sect. 5.7, we develop in detail one important application: multi-participant stereo desktop conferencing. Finally, we state conclusions in Sect. 5.8.

5.2 Multichannel Identification and the Nonuniqueness Problem

In this section, we show that the normal equation for the multichannel identification problem does not have a unique solution as in the single-channel
case. Indeed, since the P input signals are obtained by filtering from a common source, a problem of nonuniqueness is expected [121]. In the following
discussion, we suppose that the length (L) of the impulse responses (in the
transmission and receiving rooms) is equal to the length of the modeling
filters.

Assume that the system (transmission room) is linear and time invariant;
therefore, we have the following $[P(P-1)/2]$ relations [8], [19]:

$$\mathbf{x}_p^T(n)\mathbf{g}_q = \mathbf{x}_q^T(n)\mathbf{g}_p \ , \ p, q = 1, 2, ..., P \ ; \ p \neq q \ , \tag{5.1}$$

where

$$\mathbf{x}_p(n) = \left[x_p(n) \ x_p(n-1) \ \cdots \ x_p(n-L+1) \right]^T \ , \ p = 1, 2, ..., P \ ,$$

are vectors of signal samples at the microphone outputs in the transmission
room, ^{T} denotes the transpose of a vector or a matrix, and the impulse
response vectors between the source and the microphones are defined as

$$\mathbf{g}_p = \left[g_{p,0} \ g_{p,1} \ \cdots \ g_{p,L-1} \right]^T \ , \ p = 1, 2, ..., P \ .$$

Now, let us define the recursive least-squares error criterion with respect
to the modeling filters:

$$J_1(n) = \sum_{m=1}^{n} \lambda^{n-m} e^2(m) \ , \tag{5.2}$$

where λ $(0 < \lambda < 1)$ is a forgetting factor.

$$e(n) = y(n) - \sum_{p=1}^{P} \hat{\mathbf{h}}_p^T \mathbf{x}_p(n) \tag{5.3}$$

is the error signal at time n between the microphone output $y(n)$ in the
receiving room and its estimate, where

$$\hat{\mathbf{h}}_p = \left[\hat{h}_{p,0} \ \hat{h}_{p,1} \ \cdots \ \hat{h}_{p,L-1} \right]^T \ , \ p = 1, 2, ..., P \ ,$$

are the P modeling filters.

The minimization of (5.2) leads to the normal equation:

$$\mathbf{R}(n)\hat{\mathbf{h}}(n) = \mathbf{r}(n) \ , \tag{5.4}$$

where

$$\mathbf{R}(n) = \sum_{m=1}^{n} \lambda^{n-m} \mathbf{x}(m) \mathbf{x}^{T}(m)$$

$$= \begin{bmatrix} \mathbf{R}_{11}(n) & \mathbf{R}_{12}(n) & \cdots & \mathbf{R}_{1P}(n) \\ \mathbf{R}_{21}(n) & \mathbf{R}_{22}(n) & \cdots & \mathbf{R}_{2P}(n) \\ \vdots & \vdots & \ddots & \vdots \\ \mathbf{R}_{P1}(n) & \mathbf{R}_{P2}(n) & \cdots & \mathbf{R}_{PP}(n) \end{bmatrix} \tag{5.5}$$

is an estimate of the input signal covariance matrix,

$$\mathbf{r}(n) = \sum_{m=1}^{n} \lambda^{n-m} y(m) \mathbf{x}(m)$$

$$= \begin{bmatrix} \mathbf{r}_1^T(n) & \mathbf{r}_2^T(n) & \cdots & \mathbf{r}_P^T(n) \end{bmatrix}^T \tag{5.6}$$

is an estimate of the cross-correlation vector between the input and output signals (in the receiving room), and

$$\hat{\mathbf{h}}(n) = \begin{bmatrix} \hat{\mathbf{h}}_1^T(n) & \hat{\mathbf{h}}_2^T(n) & \cdots & \hat{\mathbf{h}}_P^T(n) \end{bmatrix}^T ,$$

$$\mathbf{x}(n) = \begin{bmatrix} \mathbf{x}_1^T(n) & \mathbf{x}_2^T(n) & \cdots & \mathbf{x}_P^T(n) \end{bmatrix}^T .$$

Consider the following vector:

$$\mathbf{u} = \begin{bmatrix} \sum_{p=2}^{P} \zeta_p \mathbf{g}_p^T & -\zeta_2 \mathbf{g}_1^T & \cdots & -\zeta_P \mathbf{g}_1^T \end{bmatrix}^T ,$$

where ζ_p are arbitrary factors. We can verify using (5.1) that $\mathbf{R}(n)\mathbf{u} = \mathbf{0}_{PL \times 1}$, so $\mathbf{R}(n)$ is not invertible. Vector $\mathbf{u}$ represents the nullspace of matrix $\mathbf{R}(n)$. The dimension of this nullspace depends of the number of channels and is equal to $(P - 2)L + 1$ (for $P \geq 2$). So the problem becomes worse as P increases. Thus, there is no unique solution to the problem and an adaptive algorithm will drive to any one of many possible solutions, which can be very different from the "true" desired solution $\hat{\mathbf{h}} = \mathbf{h}$. These nonunique "solutions" are dependent on the impulse responses in the transmission room:

$$\hat{\mathbf{h}}_1 = \mathbf{h}_1 + \beta \sum_{p=2}^{P} \zeta_p \mathbf{g}_p , \tag{5.7}$$

$$\hat{\mathbf{h}}_p = \mathbf{h}_p - \beta \zeta_p \mathbf{g}_1 , \quad p = 2, ..., P , \tag{5.8}$$

where β is an arbitrary factor. This, of course, is intolerable because $\mathbf{g}_p$ can change instantaneously, for example, as one person stops talking and another starts [121], [19].

5.3 The Impulse Response Tail Effect

We first define an important measure that is very useful for MCAEC.

Definition: The quantity

$$\frac{\|\mathbf{h} - \hat{\mathbf{h}}\|}{\|\mathbf{h}\|} , \qquad (5.9)$$

where $\|\cdot\|$ denotes the two-norm vector, is called the *normalized misalignment* and measures the mismatch between the impulse responses of the receiving room and the modeling filters. In the multichannel case, it is possible to have good echo cancellation even when the misalignment is large. However, in such a case, the cancellation will degrade if the $\mathbf{g}_p$ change. A main objective of MCAEC research is to avoid this problem.

Actually, for the practical case when the length of the adaptive filters is smaller than the length of the impulse responses in the transmission room, there is a unique solution to the normal equation, although the covariance matrix is very ill-conditioned.

On the other hand, we can easily show by using the classical normal equation that if the length of the adaptive filters is smaller than the length of the impulse responses in the receiving room, we introduce an important bias in the coefficients of these filters because of the strong cross-correlation between the input signals and the large condition number of the covariance matrix [19]. So in practice, we may have poor misalignment even if there is a unique solution to the normal equation.

The only way to decrease the misalignment is to partially decorrelate two-by-two the P input (loudspeaker) signals. As shown in the next section, the correlation between two channels can be linked to ill-conditioning of the covariance matrix by means of the coherence magnitude. Ill-conditioning can therefore be monitored by the coherence function which serves as a measure of achieved decorrelation. In Sect. 5.5, we summarize a number of approaches that have been developed recently for reducing the cross-correlation.

5.4 Link Between the Coherence Function and the Covariance Matrix

The covariance matrix of two concatenated stationary processes x_1 and x_2 is defined as

$$
\begin{aligned}
\mathbf{R} &= E\left\{ \begin{bmatrix} \mathbf{x}_1(n) \\ \mathbf{x}_2(n) \end{bmatrix} \begin{bmatrix} \mathbf{x}_1^T(n) \ \mathbf{x}_2^T(n) \end{bmatrix} \right\} \\
&= \begin{bmatrix} \mathbf{R}_{11} & \mathbf{R}_{12} \\ \mathbf{R}_{21} & \mathbf{R}_{22} \end{bmatrix} ,
\end{aligned} \qquad (5.10)
$$

where $E\{\cdot\}$ denotes mathematical expectation and $\mathbf{R}_{pq}$ are $L \times L$ Toeplitz matrices.

Now suppose $L \to \infty$. In this case, a Toeplitz matrix is asymptotically equivalent to a circulant matrix if its elements are absolutely summable, which is the case for the intended application. Hence, we can decompose $\mathbf{R}_{pq}$ as

$$\mathbf{R}_{pq} = \mathbf{F}^{-1}\mathbf{S}_{pq}\mathbf{F} \;,\; p, q = 1, 2 \;, \tag{5.11}$$

where $\mathbf{F}$ is the discrete Fourier transform matrix and

$$\mathbf{S}_{pq} = \mathrm{diag}\{S_{pq}(0), S_{pq}(1), \cdots, S_{pq}(L-1)\} \tag{5.12}$$

is a diagonal matrix formed by the first column of $\mathbf{F}\mathbf{R}_{pq}$. With this representation, the covariance matrix $\mathbf{R}$ can be expressed in the frequency domain as

$$\begin{aligned}
\mathbf{S} &= \begin{bmatrix} \mathbf{F} & \mathbf{0}_{L\times L} \\ \mathbf{0}_{L\times L} & \mathbf{F} \end{bmatrix} \mathbf{R} \begin{bmatrix} \mathbf{F}^{-1} & \mathbf{0}_{L\times L} \\ \mathbf{0}_{L\times L} & \mathbf{F}^{-1} \end{bmatrix} \\
&= \begin{bmatrix} \mathbf{S}_{11} & \mathbf{S}_{12} \\ \mathbf{S}_{21} & \mathbf{S}_{22} \end{bmatrix} .
\end{aligned} \tag{5.13}$$

However, since $\mathbf{S}_{pq}$ $(p, q = 1, 2)$ are diagonal matrices, the eigenvalue equation of $\mathbf{S}$ (which is the same as that of $\mathbf{R}$) has the very simple form

$$\prod_{f=0}^{L-1} \{[S_{11}(f) - \lambda_{\mathrm{s}}][S_{22}(f) - \lambda_{\mathrm{s}}] - S_{12}(f)S_{21}(f)\} = 0 \;, \tag{5.14}$$

or, assuming that $\forall f,\; S_{pp}(f) \neq 0$ $(p = 1, 2)$,

$$\prod_{f=0}^{L-1} \{S_{11}^{-1}(f)S_{22}^{-1}(f)\lambda_{\mathrm{s}}^2 - [S_{11}^{-1}(f) + S_{22}^{-1}(f)]\lambda_{\mathrm{s}} + 1 - |\gamma(f)|^2\} = 0 \;, \tag{5.15}$$

where

$$\gamma(f) = \frac{S_{12}(f)}{\sqrt{S_{11}(f)S_{22}(f)}} \tag{5.16}$$

is the coherence function between the two signals x_1 and x_2 at frequency f. Thus, the eigenvalues are obtained from L quadratics. Expression (5.15) shows that the minimum eigenvalue is lower bounded by the factor $[1 - |\gamma(f)|^2]$ and that if any $|\gamma(f)| = 1$, then the covariance matrix is singular. The eigenvalues, and hence condition number of the covariance matrix, can be obtained by trivially finding the roots of the quadratic factors.

5.5 Some Different Solutions for Decorrelation

If we have P different channels, we need to decorrelate them partially and mutually. In the following, we show how to partially decorrelate two channels.

The same process should be applied for all the channels. It is well-known that the coherence magnitude between two processes is equal to 1 if and only if they are linearly related. In order to weaken this relation, some non-linear or time-varying transformation of the stereo channels has to be made. Such a transformation reduces the coherence and hence the condition number of the covariance matrix, thereby improving the misalignment. However, the transformation has to be performed cautiously so that it is inaudible and has no effect on stereo perception.

A simple nonlinear method that gives good performance uses a half-wave rectifier [19], so that the nonlinearly transformed signal becomes

$$x_p'(n) = x_p(n) + \alpha \frac{x_p(n) + |x_p(n)|}{2} , \qquad (5.17)$$

where α is a parameter used to control the amount of nonlinearity. For this method, there can only be a linear relation between the nonlinearly transformed channels if $\forall n$, $x_1(n) \geq 0$ and $x_2(n) \geq 0$ or if we have $ax_1(n - \tau_1) = x_2(n - \tau_2)$ with $a > 0$. In practice however, these cases never occur because we always have zero-mean signals and $\mathbf{g}_1$, $\mathbf{g}_2$ are in practice never related by just a simple delay.

An improved version of this technique is to use positive and negative half-wave rectifiers on each channel respectively,

$$x_1'(n) = x_1(n) + \alpha \frac{x_1(n) + |x_1(n)|}{2} , \qquad (5.18)$$

$$x_2'(n) = x_2(n) + \alpha \frac{x_2(n) - |x_2(n)|}{2} . \qquad (5.19)$$

This principle removes the linear relation even in the special signal cases given above.

Experiments show that stereo perception is not affected by the above methods even with α as large as 0.5. Also, the distortion introduced for speech is hardly audible because of the nature of the speech signal and psychoacoustic masking effects [97]. This is explained by the following three reasons. First, the distorted signal x_p' depends only on the instantaneous value of the original signal x_p so that during periods of silence, no distortion is added. Second, the periodicity remains unchanged. Third, for voiced sounds, the harmonic structure of the signal induces "self-masking" of the harmonic distortion components. This kind of distortion is also acceptable for some music signals but may be objectionable for pure tones.

Other types of nonlinearities for decorrelating speech signals have also been investigated and compared [99]. The results indicate that, of the several nonlinearities considered, ideal half-wave rectification and smoothed half-wave rectification appear to be the best choices for speech. For music, the non-linearity parameter of the ideal rectifier must be readjusted. The smoothed rectifier does not require this readjustment but is a little more complicated to implement.

A subjectively meaningful measure to compare x_p and x'_p is not easy to find. A mathematical measure of distance, to be useful in speech processing, has to have a high correlation between its numerical value and the subjective distance judgment, as evaluated on real speech signals [109]. Since many psychoacoustic studies of perceived sound differences can be interpreted in terms of differences of spectral features, measurement of spectral distortion can be argued to be reasonable both mathematically and subjectively.

A very useful distortion measure is the Itakura-Saito (IS) measure [109], given as

$$d_{IS,p} = \frac{{a'_p}^T \mathbf{R}_{pp} a'_p}{a_p^T \mathbf{R}_{pp} a_p} - 1 \,, \tag{5.20}$$

where $\mathbf{R}_{pp}$ is the Toeplitz autocorrelation matrix of the LPC model a_p of a speech signal frame x_p and a'_p is the LPC model of the corresponding distorted speech signal frame x'_p. Many experiments in speech recognition show that if the IS measure is less than about 0.1, the two spectra that we compare are perceptually nearly identical. Simulations show that with a nonlinearity (half-wave) $\alpha = 0.5$, the IS metric is still small (about 0.03).

We could also use the ensemble interval histogram (EIH) distance (which is based on the EIH model) [57]. The interest in using this distance lies in its capability to mimic human judgement of quality. Indeed, according to [57] EIH is a very good predictor of mean opinion score (MOS), but only if the two speech observations under comparison are similar enough, which is the case here. Therefore, this measure should be a good predictor of speech signal degradation when nonlinear distortion is used. Simulations show that for a half-wave nonlinearity with $\alpha = 0.5$, the EIH distance is 1.8×10^{-3}, which is as good as a 32 kb/s ADPCM coder.

As an example, Fig. 5.1 shows the coherence magnitude between the right and left channels from the transmission room for a speech signal with $\alpha = 0$ [Fig. 5.1(a)] and $\alpha = 0.3$ [Fig. 5.1(b)]. We can see that a small amount of nonlinearity such as the half-wave rectifier reduces the coherence. As a result, the FRLS algorithm converges much better to the true solution with a reduced coherence (see Fig. 5.2).

In [112] a similar approach with non-linearities is proposed. The idea is expanded so that four adaptive filters operate on different non-linearly processed signals to estimate the echo paths. These non-linearities are chosen such that the input signals of two of the adaptive filters are independent, which thus represent a "perfect" decorrelation. Tap estimates are then copied to a fixed two-channel filter which performs the echo cancellation with the unprocessed signals. The advantage of this method is that the NLMS algorithm could be used instead of more sophisticated algorithms.

Another approach that makes it possible to use the NLMS algorithm is to decorrelate the channels by means of complementary comb filtering [121], [17]. The technique is based on removing the energy in a certain frequency

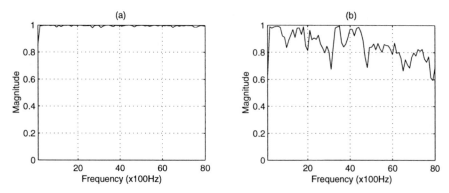

Fig. 5.1. Coherence magnitude between the right and left channels from the transmission room for a speech signal with (a) $\alpha = 0$ and (b) $\alpha = 0.3$

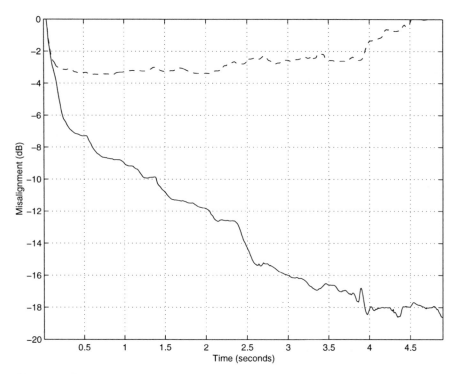

Fig. 5.2. Convergence of the misalignment using the FRLS algorithm with $\alpha = 0$ $(- -)$ and $\alpha = 0.3$ $(—)$

band of the speech signal in one channel. This means the coherence would become zero in this band and thereby results in fast alignment of the estimate even when using the NLMS algorithm. Energy is removed complementarily between the channels so that the stereo perception is not severely affected for frequencies above 1 kHz. However, this method must be combined with some other decorrelation technique for lower frequencies [20].

Two methods based on introducing time-varying filters in the transmission path were presented in [1], [71]. To show that this gives the right effect we can look at the condition for perfect cancellation (under the assumption that we do not have any zeros in the input signal spectrum) in the frequency domain which can be written as,

$$\mathcal{E}_1(\omega)G_1(\omega)T_1(\omega, n) + \mathcal{E}_2(\omega)G_2(\omega)T_2(\omega, n) = 0 , \ \forall \omega \in [-\pi, \pi] , \quad (5.21)$$

where $\mathcal{E}_p(\omega)$, $p = 1, 2$, are the Fourier transforms of the filter coefficient misalignment vectors $\boldsymbol{\varepsilon}_p = \mathbf{h}_p - \hat{\mathbf{h}}_p$. The responses $G_p(\omega)$, $p = 1, 2$, are the Fourier transformed transmission room echo paths and $T_p(\omega, n)$, $p = 1, 2$, are the frequency representation of the introduced time-varying filters. We see again for constant $T_p(\omega)$, $p = 1, 2$, the solution of (5.21) is not unique. However, for a time-varying $T_p(\omega, n)$, $p = 1, 2$, (5.21) can only be satisfied if $\mathcal{E}_p(\omega) = 0$, $\forall \omega$, $p = 1, 2$, which is the idea for this method. In [1], left and right signals are filtered through two independent time-varying first-order all-pass filters. Stochastic time-variation is introduced by making the pole position of the filter a random walk process. The actual position is limited by the constraints of stability and inaudibility of the introduced distortion. While significant reduction in correlation can be achieved for higher frequencies with the imposed constraints, the lower frequencies are still fairly unaffected by the time-variation. In [71], a periodically varying filter is applied to one channel so that the signal is either delayed by one sample or passed through without delay. A transition zone between the delayed and non-delayed state is also employed in order to reduce audible discontinuities. This method may also affect the stereo perception.

Although the idea of adding independent perceptually shaped noise to the channels was mentioned in [121], [19], thorough investigations of the actual benefit of the technique was not presented. Results regarding variants of this idea can be found in [50], [58]. A pre-processing unit estimating the masking threshold and adding an appropriate amount of noise was proposed in [58]. It was also noted that adding a masked noise to each channel may affect the spatialization of the sound even if the noise is inaudible at each channel separately. This effect can be controlled through correction of the masking threshold when appropriate. In [50], the improvement of misalignment was studied in the SAEC when a perceptual audio coder was added in the transmission path. Reduced correlation between the channels was shown by means of coherence analysis, and improved convergence rate of the adaptive algorithm was observed. A low-complexity method for achieving additional decorrelation by modifying the decoder was also proposed. The encoder cannot quantize every single frequency band optimally due to rate constraints. This has the effect that there is a margin on the masking threshold which can be exploited. In the presented method, the masking threshold is estimated from the modified discrete cosine transform (MDCT) coefficients delivered by the encoder, and an appropriate inaudible amount of decorrelating noise is added to the signals.

In the rest of this chapter, we suppose that one of the previous decorrelation methods is used so the normal equation has a unique solution. However, the input signals can still be highly correlated, therefore requiring special treatment.

5.6 Conventional Multichannel Time-Domain Adaptive Filters

In this section, we derive three important classical adaptive algorithms: the multichannel RLS algorithm which is optimal from a convergence point of view, the multichannel LMS algorithm which is simple to implement but converges slowly in general, and finally the multichannel APA.

5.6.1 The Multichannel RLS Algorithm

From the normal equation (5.4), we easily derive the classical update equations for the multichannel RLS:

$$e(n) = y(n) - \hat{\mathbf{h}}^T(n - 1)\mathbf{x}(n) , \tag{5.22}$$

$$\hat{\mathbf{h}}(n) = \hat{\mathbf{h}}(n - 1) + \mathbf{R}^{-1}(n)\mathbf{x}(n)e(n) . \tag{5.23}$$

Using the matrix inversion lemma, we obtain the following recursive equation for the inverse of the covariance matrix:

$$\mathbf{R}^{-1}(n) = \lambda^{-1}\mathbf{R}^{-1}(n - 1)$$
$$- \frac{\lambda^{-2}\mathbf{R}^{-1}(n - 1)\mathbf{x}(n)\mathbf{x}^T(n)\mathbf{R}^{-1}(n - 1)}{1 + \lambda^{-1}\mathbf{x}^T(n)\mathbf{R}^{-1}(n - 1)\mathbf{x}(n)} . \tag{5.24}$$

Because RLS has so far proven to perform better than other algorithms in the MCAEC application [5], a fast calculation scheme for a multichannel version is presented in Chap. 6. Compared to standard RLS, fast RLS has a lower complexity, $6P^2L + 2PL$ multiplications [instead of $O(P^2L^2)$]. This algorithm is a numerically stabilized version of the algorithm proposed in [4]. Some extra stability control has to be added so that the algorithm behaves well for a non-stationary speech signal.

5.6.2 The Multichannel LMS Algorithm

The mean-square error criterion is defined as

$$J_2 = E\left\{\left[y(n) - \hat{\mathbf{h}}^T\mathbf{x}(n)\right]^2\right\} . \tag{5.25}$$

Let $\nabla J_2(\hat{\mathbf{h}})$ denote the value of the gradient with respect to $\hat{\mathbf{h}}$. According to the steepest descent method, the updated value of $\hat{\mathbf{h}}$ at time n is computed by using the simple recursive relation [66]:

$$\hat{\mathbf{h}}(n) = \hat{\mathbf{h}}(n-1) - \frac{\mu}{2}\nabla J_2[\hat{\mathbf{h}}(n-1)] , \qquad (5.26)$$

where μ is a positive step-size constant. Differentiating (5.25) with respect to the filter, we get the following value for the gradient:

$$\nabla J_2(\hat{\mathbf{h}}) = \partial J_2/\partial \hat{\mathbf{h}}$$
$$= -2\mathbf{r} + 2\mathbf{R}\hat{\mathbf{h}} , \qquad (5.27)$$

with $\mathbf{r} = E\{y(n)\mathbf{x}(n)\}$ and $\mathbf{R} = E\{\mathbf{x}(n)\mathbf{x}^T(n)\}$. By taking $\nabla J_2(\hat{\mathbf{h}}) = \mathbf{0}_{LP\times 1}$, we obtain the Wiener-Hopf equation

$$\mathbf{R}\hat{\mathbf{h}} = \mathbf{r} , \qquad (5.28)$$

which is similar to the normal equation (5.4) that was derived from a weighted least-squares criterion (5.2). Note that we use the same notation for similar variables derived from either the Wiener-Hopf equation or the normal equation.

The steepest-descent algorithm is now:

$$\hat{\mathbf{h}}(n) = \hat{\mathbf{h}}(n-1) + \mu E\{\mathbf{x}(n)e(n)\} , \qquad (5.29)$$

and the classical stochastic approximation (consisting of approximating the gradient with its instantaneous value) [66] provides the multichannel LMS algorithm,

$$\hat{\mathbf{h}}(n) = \hat{\mathbf{h}}(n-1) + \mu\mathbf{x}(n)e(n) , \qquad (5.30)$$

of which the classical mean weight convergence condition under appropriate independence assumption is:

$$0 < \mu < \frac{2}{L\sum_{p=1}^{P}\sigma_{x_p}^2} , \qquad (5.31)$$

where the $\sigma_{x_p}^2$ $(p = 1, 2, ..., P)$ are the powers of the input signals. When this condition is satisfied, the weight vector converges in the mean to the optimal Wiener-Hopf solution.

5.6.3 The Multichannel APA

The affine projection algorithm (APA) [106] has become popular because of its lower complexity, compared to RLS, while it converges almost as fast. Therefore, it is interesting to derive and study the multichannel version of this algorithm.

The Straightforward Multichannel APA

A simple way for obtaining the single-channel APA is to search for an algorithm of the stochastic gradient type cancelling N *a posteriori* errors [96]. This requirement results in an underdetermined set of linear equations of which the mininum-norm solution is chosen. In the following, this technique is extended in order to fit our problem [9].

By definition, the set of N *a priori* errors and N *a posteriori* errors are:

$$\mathbf{e}(n) = \mathbf{y}(n) - \mathbf{X}^T(n)\hat{\mathbf{h}}(n-1) , \tag{5.32}$$

$$\mathbf{e}_a(n) = \mathbf{y}(n) - \mathbf{X}^T(n)\hat{\mathbf{h}}(n) , \tag{5.33}$$

where

$$\mathbf{X}(n) = \left[\mathbf{X}_1^T(n)\, \mathbf{X}_2^T(n) \cdots \mathbf{X}_P^T(n) \right]^T$$

is a matrix of size $PL \times N$ and the $L \times N$ sub-matrices

$$\mathbf{X}_p(n) = \left[\mathbf{x}_p(n)\, \mathbf{x}_p(n-1) \cdots \mathbf{x}_p(n-N+1) \right]$$

are made from the N last input vectors $\mathbf{x}_p(n)$; finally, $\mathbf{y}(n)$ and $\mathbf{e}(n)$ are respectively vectors of the N last samples of the reference signal $y(n)$ and error signal $e(n)$.

Using (5.32) and (5.33) plus the requirement that $\mathbf{e}_a(n) = \mathbf{0}_{N\times 1}$, we obtain:

$$\mathbf{X}^T(n)\Delta\hat{\mathbf{h}}(n) = \mathbf{e}(n) , \tag{5.34}$$

where $\Delta\hat{\mathbf{h}}(n) = \hat{\mathbf{h}}(n) - \hat{\mathbf{h}}(n-1)$.

Equation (5.34) (N equations in PL unknowns, $N \leq PL$) is an underdetermined set of linear equations. Hence, it has an infinite number of solutions, out of which the minimum-norm solution is chosen, so that the adaptive filter has smooth variations. This results in [9], [113]:

$$\hat{\mathbf{h}}(n) = \hat{\mathbf{h}}(n-1) + \mathbf{X}(n) \left[\mathbf{X}^T(n)\mathbf{X}(n) \right]^{-1} \mathbf{e}(n) . \tag{5.35}$$

However, in this straightforward APA, the normalization matrix $\mathbf{X}^T(n)\mathbf{X}(n) = \sum_{p=1}^{P} \mathbf{X}_p^T(n)\mathbf{X}_p(n)$ does not involve the cross-correlation elements of the P input signals [namely $\mathbf{X}_p^T(n)\mathbf{X}_q(n)$, $p,q = 1,2,...,P$, $p \neq q$] and this algorithm may converge slowly.

The Improved Two-Channel APA

A simple way to improve the previous adaptive algorithm is to use some othogonality properties, which will be shown later to appear in this context. We use the constraint that $\Delta\hat{\mathbf{h}}_p$ be orthogonal to $\mathbf{X}_q$, $p \neq q$. As a result, we take into account separately the contributions of each input signal. These constraints read:

$$\mathbf{X}_2^T(n)\Delta\hat{\mathbf{h}}_1(n) = \mathbf{0}_{N\times 1} \, , \tag{5.36}$$

$$\mathbf{X}_1^T(n)\Delta\hat{\mathbf{h}}_2(n) = \mathbf{0}_{N\times 1} \, , \tag{5.37}$$

and the new set of linear equations characterizing the improved two-channel APA is:

$$\begin{bmatrix} \mathbf{X}_1^T(n) \, \mathbf{X}_2^T(n) \\ \mathbf{X}_2^T(n) \, \mathbf{0}_{N\times L} \\ \mathbf{0}_{N\times L} \, \mathbf{X}_1^T(n) \end{bmatrix} \begin{bmatrix} \Delta\hat{\mathbf{h}}_1(n) \\ \Delta\hat{\mathbf{h}}_2(n) \end{bmatrix} = \begin{bmatrix} \mathbf{e}(n) \\ \mathbf{0}_{N\times 1} \\ \mathbf{0}_{N\times 1} \end{bmatrix} \, . \tag{5.38}$$

The improved two-channel APA algorithm is given by the minimum-norm solution of (5.38) which is found as [9],

$$\Delta\hat{\mathbf{h}}_1(n) = \mathbf{Z}_1(n) \left[\mathbf{Z}_1^T(n)\mathbf{Z}_1(n) + \mathbf{Z}_2^T(n)\mathbf{Z}_2(n) \right]^{-1} \mathbf{e}(n) \, , \tag{5.39}$$

$$\Delta\hat{\mathbf{h}}_2(n) = \mathbf{Z}_2(n) \left[\mathbf{Z}_1^T(n)\mathbf{Z}_1(n) + \mathbf{Z}_2^T(n)\mathbf{Z}_2(n) \right]^{-1} \mathbf{e}(n) \, , \tag{5.40}$$

where $\mathbf{Z}_p(n)$ is the projection of $\mathbf{X}_p(n)$ onto a subspace orthogonal to $\mathbf{X}_q(n)$, $p \neq q$, i.e.,

$$\mathbf{Z}_p(n) = \left\{ \mathbf{I}_{L\times L} - \mathbf{X}_q(n) \left[\mathbf{X}_q^T(n)\mathbf{X}_q(n) \right]^{-1} \mathbf{X}_q(n) \right\} \mathbf{X}_p(n) \, , \tag{5.41}$$

$$p,q = 1,2 \, , \ p \neq q \, .$$

This results in the following orthogonality conditions,

$$\mathbf{X}_p^T(n)\mathbf{Z}_q(n) = \mathbf{0}_{N\times N} \, , \ p \neq q \, . \tag{5.42}$$

The Improved Multichannel APA

The algorithm explained for two channels is easily generalized to an arbitrary number of channels P. Define the following (multichannel) matrices of size $L \times (P-1)N$:

$$\mathbf{X}_{\mathrm{m},p}(n) = \begin{bmatrix} \mathbf{X}_1(n) \, \cdots \, \mathbf{X}_{p-1}(n) \, \mathbf{X}_{p+1}(n) \, \cdots \, \mathbf{X}_P(n) \end{bmatrix} \, ,$$

$$p = 1,2,...,P \, .$$

The P orthogonality constraints are:

$$\mathbf{X}_{\mathrm{m},p}^T(n)\Delta\hat{\mathbf{h}}_p(n) = \mathbf{0}_{(P-1)N\times 1} \, , \ p = 1,2,...,P \, , \tag{5.43}$$

and by using the same steps as for $P = 2$, a solution similar to (5.39), (5.40) is obtained [9]:

$$\Delta\hat{\mathbf{h}}_p(n) = \mathbf{Z}_p(n) \left[\sum_{q=1}^{P} \mathbf{Z}_q^T(n)\mathbf{Z}_q(n) \right]^{-1} \mathbf{e}(n) \, , \ p = 1,2,...,P \, , \tag{5.44}$$

where $\mathbf{Z}_p(n)$ is the projection of $\mathbf{X}_p(n)$ onto a subspace orthogonal to $\mathbf{X}_{\mathrm{m},p}(n)$, i.e.,

$$\mathbf{Z}_p(n) = \left\{ \mathbf{I}_{L \times L} - \mathbf{X}_{\mathrm{m},p}(n) \left[\mathbf{X}_{\mathrm{m},p}^T(n) \mathbf{X}_{\mathrm{m},p}(n) \right]^{-1} \mathbf{X}_{\mathrm{m},p}(n) \right\} \mathbf{X}_p(n) \ ,$$

$$p = 1, 2, ..., P \ . \tag{5.45}$$

Note that this equation holds only under the condition $L \geq (P-1)N$, so that the matrix that appears in (5.45) is invertible.

We can easily see that:

$$\mathbf{X}_{\mathrm{m},p}^T(n) \mathbf{Z}_p(n) = \mathbf{0}_{(P-1)N \times N} \ , \quad p = 1, 2, ..., P \ . \tag{5.46}$$

5.7 Application: Synthesized Stereo and AEC for Desktop Conferencing

In this section, we focus on one particular application: *multi-participant stereo desktop conferencing* [18]. With single-channel sound, simultaneous talkers are overlaid and it is difficult to concentrate on one particular voice. On the other hand, by using our binaural auditory system together with multichannel presentation, we can concentrate on one source to the exclusion of others (the so-called cocktail party effect). Moreover, localization helps us identify which person is actually talking. This is a very difficult task in a mono presentation. Communication with stereo (or multichannel) sound likely will grow rapidly in the near future, especially over the Internet.

The general scenario is as follows. Several persons in different locations would like to communicate with each other, and each one has a workstation. Each participant would like to see on the screen pictures of the other participants arranged in a reasonable fashion and to hear them in perceptual space in a way that facilitates identification and understanding. For example, the voice of a participant whose picture is located on the left of the screen should appear to come from the left.

We suppose that we are located in a hands-free environment, where the composite acoustic signal is presented over loudspeakers. This study will be limited to two channels, so we assume that each workstation is equipped with two loudspeakers (one on each side of the screen) and one microphone (somewhere on top of the screen, for example). As we will see later, a very convenient method using two loudspeakers can accomodate up to four participants. This arrangement can be generalized to create more images. However, it is not clear how many images a participant can conveniently deal with.

Figure 5.3 shows the configuration for a microphone at the local site, where h_1 and h_2 represent the two echo paths between the two loudspeakers and the microphone. The two reference signals x_1 and x_2 from the remote sites are obtained by synthesizing stereo sound from the outputs of all the remote single microphones. The nonuniqueness arises because, for each remote site, the signals are derived by filtering from a common source.

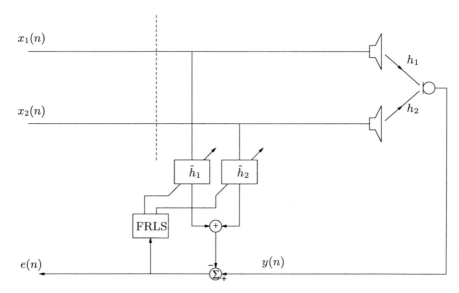

Fig. 5.3. Schematic diagram of stereophonic echo cancellation for desktop conferencing

5.7.1 Interchannel Differences for Synthesizing Stereo Sound

In the following scenario, we assume that two loudspeakers are positioned symmetrically on each side of the screen and that the conferee is in front of the screen, close to and approximately centered between the loudspeakers. The location of auditory images in perceptual space is controlled by interchannel intensity and time differences and is mediated by the binaural auditory system.

In any discussion of the relationship between interchannel differences and perceptual effects, it is important to maintain a clear distinction between *interchannel* and *interaural* differences. If sounds are presented to the two ears by means of headphones, the interaural intensity and time differences ΔI_a and $\Delta \tau_a$ can be controlled directly. If signals are presented over a pair of loudspeakers, each ear receives both the left- and right-channel signals. The left-channel signal arrives earlier and is more intense at the left ear than at the right, and vice versa, so that interchannel intensity and time differences ΔI_c and $\Delta \tau_c$ influence ΔI_a and $\Delta \tau_a$, but in general interaural intensity and time differences cannot be controlled directly. In addition to perceptual effects produced by interaural time and intensity differences, localization of sounds presented over a pair of loudspeakers is also influenced by the *precedence effect* [53]: when identical or nearly identical sounds come to a listener from two loudspeakers, the sound appears to originate at the loudspeaker from which the sound arrives first.

To arrange the acoustic images, we can manipulate interchannel intensity and time differences either separately or together. If two identical signals are presented to the two loudspeakers, so that there are no interaural differences, the image will be well fused and localized in the median plane. As the interchannel intensity ratio varies from unity, the image will move toward the loudspeaker receiving the more intense signal. If, instead, the interchannel time difference is varied, the image will in general move toward the loudspeaker receiving the leading signal [36], [135].

Pure Interchannel Intensity Difference

It is well known that the effect of introducing an interchannel intensity ratio ΔI_c into signals that are otherwise identical is to move the image away from the median plane toward the loudspeaker receiving the more intense signal. Recent experiments conducted in our laboratory for a desktop configuration, as well as previous experiments with conventional loudspeaker placement in a room [22], indicate that a 20-dB interchannel intensity ratio produces almost complete lateralization.

If there are two remote conferees, experiments with headphones conducted in our laboratory suggest that interchannel intensity difference may be the best choice for desktop conferencing in terms of auditory localization and signal separation. The suggested strategy is to present the acoustic signal from one remote participant to one loudspeaker and the acoustic signal from the other remote participant to the other loudspeaker. With three remote participants, the suggested strategy would be the same for the first two participants with the acoustic signal from the third remote participant presented equally to both loudspeakers. Thus, communication with good localization and signal separation among four conferees (one local plus three remote) appears to be feasible. The number of participants could be increased by using finer gradations of ΔI_c, but separating the different remote talkers would be more difficult.

Pure Interchannel Time Difference

The nature of the signal plays a more important role in localization and signal separation for interchannel time difference $\Delta \tau_c$ than for interchannel intensity difference ΔI_c. For pure tones, the binaural system is insensitive to interaural time differences $\Delta \tau_a$ for frequencies substantially above 1.5 kHz [36], and for lower frequencies, localization of the image is periodic in the interaural time difference $\Delta \tau_a$, with a period equal to the period of the tone. For complex signals with some envelope structure, localization is influenced by both low- and high-frequency interaural time differences. Since, as discussed above, the interaural time difference $\Delta \tau_a$ is indirectly influenced by interchannel time difference $\Delta \tau_c$, it follows that the nature of the signal plays an important role in localization and signal separation for interchannel time difference $\Delta \tau_c$.

If there are two remote talkers, a suggested strategy for localization with interchannel time difference is to present the acoustic signal from one remote participant with an interchannel time difference $\Delta\tau_c = 1$ ms and the acoustic signal from the other remote participant with an interchannel time difference $\Delta\tau_c = -1$ ms. It has been known for a long time [36] that, with headphone presentation, lateralization increases only slightly as interaural time difference increases above 1 ms. Recent experiments with desktop loudspeakers, as well as previous experiments with conventional loudspeaker placement in a room [22], show much the same effect. We do not know how interchannel time difference specifically affects the cocktail party effect.

Combined Interchannel Intensity and Time Differences

As discussed in the previous two subsections, the localization of a sound image can be influenced by both the interchannel intensity difference ΔI_c and the interchannel time difference $\Delta\tau_c$. To a certain extent, and within limits, these two types of interchannel differences are tradable in the sense that the same localization can be achieved with various combinations of the two variables. For example, one can achieve roughly the same image position with an amplitude shift, a time shift, or an appropriate combination of the two (time-intensity trading). Furthermore, under some conditions, intensity difference and time difference can be used to reinforce each other to provide a larger shift than is achievable by either one alone.

5.7.2 Choice of Interchannel Differences for Stereo AEC

In principle, for localization with three remote talkers, the best choice of interchannel difference is ΔI_c. But if we want to synthesize a remote talker on the right (resp. left), speech energy will be present only on the right (resp. left) loudspeaker, so we will be able to identify only one impulse response (from this loudspeaker to the microphone) and not the other one. From an acoustic echo cancellation point of view this situation is highly undesirable. For example, if the remote talker on the right stops talking and the remote talker on the left begins, the adaptive algorithm will have to reconverge to the corresponding acoustic path because, in the meantime, it will have "forgotten" the other acoustic path. Therefore, the adaptive algorithm will have to track the different talkers continually, reconverging for each one, so the system will become uncontrollable — especially in a nonstationary environment (changes of the acoustic paths) and in double-talk situations. As a result, we will have degraded echo cancellation much of the time.

The solution to this problem is that, for each remote talker, we must have some energy on both loudspeakers to be able to maintain identification of the two impulse responses between loudspeakers and the microphone. Thus, the optimal choice of interchannel difference from an acoustic echo cancellation point of view is pure $\Delta\tau_c$ since energy is equally presented to both loudspeakers for all remote talkers. However, in practice, this choice may not be enough for good localization. Therefore, combined $\Delta I_c / \Delta \tau_c$ seems to be the best compromise between good localization and echo cancellation.

If there are two remote talkers, a strategy for good localization and echo cancellation would be to present the acoustic signal from one remote participant to both loudspeakers with $\Delta I_c = 6\,\mathrm{dB}$, $\Delta \tau_c = 1\,\mathrm{ms}$ and the acoustic signal from the other remote participant to both loudspeakers with $\Delta I_c = -6\,\mathrm{dB}$, $\Delta \tau_c = -1\,\mathrm{ms}$. With three remote participants, the suggested strategy would be the same for the first two participants with the addition of the third remote participant's microphone signal presented to both loudspeakers with $\Delta I_c = 0\,\mathrm{dB}$, $\Delta \tau_c = 0\,\mathrm{ms}$.

Thus, for any remote participant's microphone signal s_j, the contribution to the local synthesized stereo signals is written

$$x_i(n) = g_{j,i}(n) * s_j(n) , \quad i = 1,2 , \tag{5.47}$$

where $*$ denotes convolution and $g_{j,i}$ are the impulse responses for realizing the desired $\Delta I_c / \Delta \tau_c$. For example, with the above suggested $6\,\mathrm{dB}$, $1\,\mathrm{ms}$ values, a talker on the left, say, would be synthesized with

$$\mathbf{g}_{j,1} = \begin{bmatrix} 1 & 0 & \cdots & 0 & 0 \end{bmatrix}^T ,$$
$$\mathbf{g}_{j,2} = \begin{bmatrix} 0 & 0 & \cdots & 0 & 0.5 \end{bmatrix}^T ,$$

where the number of samples of delay in $g_{j,2}$ corresponds to $1\,\mathrm{ms}$.

Figure 5.4 shows how the signals from N remote conferees are combined to produce the local synthesized signals x_1, x_2. Each $g_{j,1}$, $g_{j,2}$ pair is selected as exemplified above to locate the acoustic image in some desired position. There is some flexibility as to where the synthesis function is located and the most efficient deployment will depend on the particular system architecture.

A composite diagram of the synthesis, nonlinear transformation, and stereo AEC appears in Fig. 5.5. This shows the complete local signal processing suite for one conferee. This same setup is reproduced for each conferee, making obvious permutations on the remote signals and choice of synthesis filters. Further details and simulation results can be found in [18].

5.7.3 Simulations

First, we wish to show, by way of simulation, the convergence of the adaptive algorithm without and with a nonlinearity when a remote conferee talks for a while, stops talking, and another remote conferee starts to talk. To better show this effect, we use two independent white noises as the two remote sources and we synthesize the stereo effect using $\Delta I_c = 6\,\mathrm{dB}$, $\Delta \tau_c = 1\,\mathrm{ms}$ for Source 1 and $\Delta I_c = -6\,\mathrm{dB}$, $\Delta \tau_c = -1\,\mathrm{ms}$ for Source 2. The microphone output signal (y) in the receiving room is obtained by summing the two convolutions $(h_1 * x_1)$ and $(h_2 * x_2)$, where h_1 and h_2 were measured in an

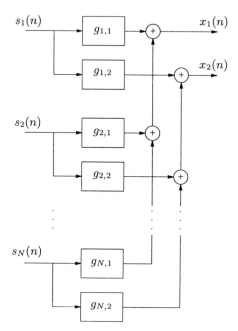

Fig. 5.4. Synthesizing local stereo signals from N remote signals

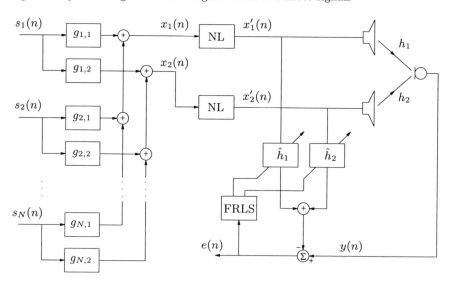

Fig. 5.5. Overall diagram of synthesis, nonlinear transformation, and stereo acoustic echo cancellation

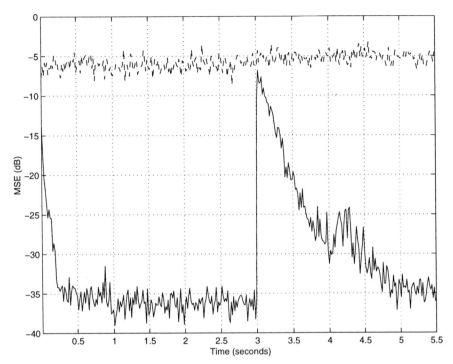

Fig. 5.6. Behavior of the MSE (solid line) without nonlinearity ($\alpha = 0$). Sources are two independent white noises and stereo sound was synthesized differently for the two sources (dashed line). Source 1 talks first for 3 s. At time 3 s, Source 1 stops talking and Source 2 starts to talk for 2.5 s

actual room as 4096-point responses. The length of the adaptive filters $\hat{h}_1$ and $\hat{h}_2$ is taken as $L = 1000$. An additional white noise with 40 dB ENR (echo-to-noise ratio) is added to the microphone signal $y(n)$. The sampling frequency is 8 kHz. For all of our simulations, we have used the two-channel FRLS algorithm [5]. Source 1 talks first for 3 s. At time 3 s, Source 1 stops talking and Source 2 starts to talk for 2.5 s. Figure 5.6 shows the behavior of the mean-square error (MSE) without a nonlinearity while Fig. 5.7 shows the behavior of the MSE with a nonlinearity (positive and negative half-wave, $\alpha = 0.5$). For the purpose of smoothing the curves, error samples are averaged over 128 points. We can see that the adaptive algorithm is very sensitive to changes in the remote talkers without a nonlinearity. When changes occur, the algorithm diverges and has to reconverge again. However, with a nonlinearity, the adaptive algorithm is much more robust.

In the second part of these simulations, we show the effectiveness of our nonlinear transformation method on the coherence and misalignment with synthetic stereo sound using actual speech signals. A stereo signal was syn-

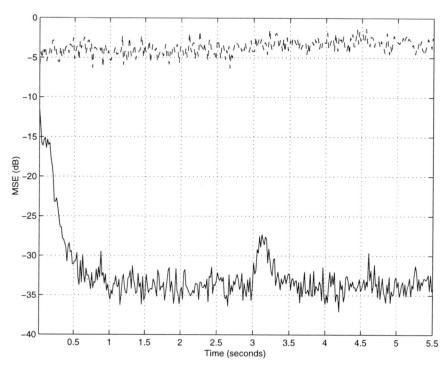

Fig. 5.7. Behavior of the MSE (solid line) with nonlinearity (positive and negative half-wave, $\alpha = 0.5$). Other conditions same as in Fig. 5.6

thesized with $\Delta I_c = 0\,\mathrm{dB}$, $\Delta \tau_c = 0\,\mathrm{ms}$, using speech signal sampled at $8\,\mathrm{kHz}$. So, the two signals x_1 and x_2 presented at the loudspeakers in the receiving room are identical (this is the most difficult example). Here we use again the positive and negative half-wave rectifiers for x_1 and x_2 with $\alpha = 0.5$. With this value, there is no audible degradation of the original signal. Everything else is the same as in the previous example. Figure 5.8 shows the behavior of the coherence magnitude and the misalignment (for the former, we use MATLAB's spectrum function and take the square root of the computed magnitude-squared coherence). The coherence magnitude samples are averaged over 100 points. We note that the coherence is reduced, which implies a great improvement of the misalignment. Without the nonlinear transformation, the signals x_1 and x_2 are identical, so the coherence magnitude is equal to 1 at all frequencies and the misalignment is not much less than unity ($0\,\mathrm{dB}$).

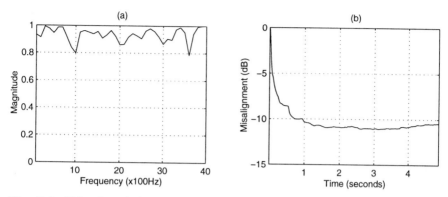

Fig. 5.8. Behavior of the coherence magnitude (a) and misalignment (b) with synthetic stereo sound and $\alpha = 0.5$

5.8 Conclusions

In this chapter, we have given an overview on multichannel acoustic echo cancellation. We have focused on presenting the main theoretical results that describe and explain the important difference between monophonic and multichannel echo cancellation, and have presented classical multichannel time-domain adaptive filtering algorithms. The multichannel echo canceler has to track echo path changes in both transmission and receiving rooms unless the transmitted multichannel signals are modified so that the normal equation has a unique solution. The sensitivity can only be reduced by decreasing the interchannel coherence; modifications of the adaptive algorithm itself will not decrease sensitivity. Obviously, there is a limit to the decrease of this interchannel coherence; a strong decorrelation will destroy the stereo effect. Because of the high correlation among the input signals (even after being processed), adaptive algorithms that take this correlation into account should be used, in order to have good performance in terms of convergence rate and tracking of the MCAEC. We have also talked about the multi-participant desktop conferencing application, presenting the most useful ways to synthesize stereo and showing how they can be combined with stereophonic acoustic echo cancellation.

6. A Fast Normalized Cross-Correlation DTD Combined with a Robust Multichannel Fast Recursive Least-Squares Algorithm

6.1 Introduction

Ideally, acoustic echo cancelers (AECs) remove undesired echoes that result from acoustic coupling between the loudspeaker and the microphone used in full-duplex hands-free telecommunication systems. Figure 6.1 shows a diagram of a single-channel AEC. The far-end speech signal $x(n)$ goes through the echo path represented by a filter $h(n)$ to produce the echo, $y_e(n)$, which is picked up by the microphone together with the near-end talker signal $v(n)$ and ambient noise $w(n)$. The composite microphone signal is denoted $y(n)$.

The echo path is modeled by an adaptive finite impulse response filter, $\hat{h}(n)$, which subtracts a replica of the echo from the return channel and thereby achieves cancellation. This may look like a simple straightforward system identification task for the adaptive filter. However, in most conversations there are the so-called *double-talk* situations that make identification much more problematic than what it might appear at first glance. Double-

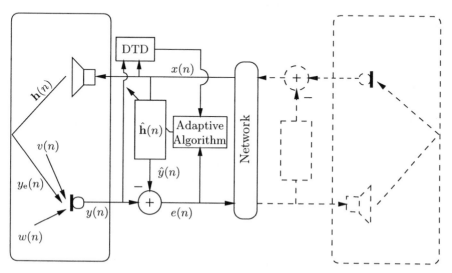

Fig. 6.1. Block diagram of a basic AEC setup and double-talk detector

talk occurs when the speech of the two talkers arrives simultaneously at the canceler, i.e. $x(n) \neq 0$ and $v(n) \neq 0$.

In the double-talk situation, the near-end speech acts as a strong uncorrelated noise to the adaptive algorithm. The disturbing near-end speech may cause the adaptive filter to diverge, allowing annoying audible echo to pass through to the far-end. The usual way to alleviate this problem is to slow down or completely halt the filter adaptation when near-end speech is detected. This is the very important role of the double-talk detector (DTD). The generic double-talk detection scheme is based on a comparison of a detection statistic, ξ, to a preset threshold T. An undisputable fact, however, is that all double-talk detectors have a probability of not detecting double-talk when it occurs, i.e., they have a probability of miss, $P_m \neq 0$. Reducing P_m (by adjusting the threshold T) will increase the probability of false alarm (P_f) and hence slow down the convergence rate of the acoustic echo canceler. Also, requiring a small probability of miss increases the sensitivity of the DTD to echo path changes. Therefore, the value of P_m cannot be set too small. As a consequence, no matter what DTD is used, undetected near-end speech will perturb the adaptive algorithm from time to time. In practice, what has been done in the past is as follows: first, the DTD is designed to be "as good as" one can afford, and then the adaptive algorithm is slowed down so that it does not significantly diverge due to the detection errors made by the DTD. This is natural to do since, if the adaptive algorithm is very fast, it can react faster to changes (e.g. double-talk) than can the DTD, and thus diverge. However, this approach severely penalizes the convergence rate of the AEC in the ideal situation, i.e. when far-end, but no near-end, talk is present.

In the light of these facts, we look at adaptive algorithms that can handle at least a small amount of double-talk without diverging. This is achieved by combining the following three concepts which will be studied in this chapter:

- adaptive algorithms,
- robust statistics, and
- double-talk detection schemes.

Many DTD schemes have been proposed since the echo canceler was invented [117]. The Geigel algorithm [37] has proven successful in network echo cancelers; however, it does not always provide reliable performance when used in AECs. This is because its assumption of a minimum echo path attenuation may not be valid in the acoustic case. Other methods based on cross-correlation and coherence [140], [52], [16] have been studied which appear to be more appropriate for AEC applications. A DTD based on multi-statistic testing in combination with modeling of the echo path by two filters was proposed in [104].

In this chapter, we will present a method for handling double-talk that uses a robust fast recursive least-squares (FRLS) algorithm and the normalized cross-correlation (NCC) DTD [16]. The NCC DTD is developed into a fast version, called FNCC, by reusing computational results from the FRLS algorithm. The major advantage of this detector is that it is not dependent on the actual echo path attenuation the way, e.g., the Geigel DTD is. This makes it much more suitable for the acoustic application. This NCC detector is also shown to be a Neyman-Pearson detector [111] under certain assumptions. Furthermore, as a by-product of the fast NCC derivation, we show that the FNCC DTD uses the same calculation structure, but not the same test statistic, as the two-path method proposed in [104]. Hence, our derivation theoretically justifies the two-path structure.

The chapter is organized as follows. Section 6.2 presents the generic double-talk detector and describes the properties of DTDs and adaptive algorithms. We devote Sect. 6.3 to robust recursive least-squares algorithms since the combination of robust algorithms and double-talk detectors have been found successful. Furthermore, the recursive least-squares algorithms are especially suited for multichannel acoustic echo cancellation. Section 6.4 reviews the NCC DTD. A fast version of this algorithm (FNCC) is derived and a link is established between this algorithm and the two-path structure. Section 6.5 presents some practical modifications of the FNCC DTD and FRLS echo canceler in order to improve over-all performance. Evaluation of the proposed FNCC algorithm is made in Sect. 6.6. We also present simulations showing the effectiveness of the DTD and an FRLS-based acoustic echo canceler. Section 6.7 ends this chapter with conclusions and remarks for further studies.

6.2 The Generic Double-Talk Detector Scheme

All types of double-talk detectors basically operate in the same manner. Thus, the general procedure for handling double-talk is described by the following four steps, [28]:

1. A detection statistic, ξ, is formed using available signals, e.g. x, y, e, etc., and the estimated filter coefficients $\hat{h}$.
2. The detection statistic ξ is compared to a preset threshold T, and double-talk is declared if $\xi < T$. (Although in general detection theory detection is declared when the detection statistic *exceeds* the threshold, we choose the opposite convention here for convenience in our particular problem.)
3. Once double-talk is detected, it is declared for a minimum period of time t_{hold}. During periods for which double-talk is declared, the filter adaptation is disabled.
4. If $\xi \geq T$ continuously for an interval of t_{hold} seconds, the filter resumes adaptation. The comparison of ξ to T continues, and double-talk is declared again when $\xi < T$.

The hold time t_{hold} in Step 3 and Step 4 may be necessary to suppress detection dropouts due to the noisy behavior of the detection statistic (i.e. if an inaccurate detection statistics is used). Although there are some possible variations, most of the DTD algorithms have this basic form and differ only in the detection statistic.

An "optimum" decision variable ξ for double-talk detection should behave as follows:

- If $v(n) = 0$ (double-talk is not present), $\xi \geq T$,
- If $v(n) \neq 0$ (double-talk is present), $\xi < T$,
- ξ is insensitive to echo path variations.

The threshold T must be a constant, independent of the data. Moreover, it is desirable that the decisions are made without introducing undue delay; delayed decisions adversely affect the performance of the AEC because, in order to prevent divergence, either the model filter update must be delayed or the adaptation slowed down.

6.2.1 Properties of DTDs and Adaptive Algorithms

Important issues that have to be addressed when designing a DTD and an echo canceler are:

(i)   Under what circumstances does double-talk disturb the adaptive filter?
(ii)   What characterizes "good" double-talk detectors?
(iii) In what situation does the DTD operate well/poorly?

The performance of an AEC is often evaluated through its mean-square error (MSE) performance or through its misalignment $\|\boldsymbol{\varepsilon}(n)\|^2 = \|\mathbf{h} - \hat{\mathbf{h}}(n)\|^2$ in different situations. The expected *normalized* misalignment of the RLS algorithm is given by the formula (Sect. 9.4.1):

$$\frac{E\{\|\boldsymbol{\varepsilon}(n)\|^2\}}{\|\mathbf{h}\|^2} \approx \frac{(1-\lambda)L}{2\,\mathrm{EBR}} \ . \tag{6.1}$$

Here $E\{\cdot\}$ denotes mathematical expectation, λ is a constant parameter of the recursive least-squares adaptive algorithm, L is the length of the adaptive filter and

$$\mathrm{EBR} = \frac{\sigma_{y_e}^2}{\sigma_v^2 + \sigma_w^2} \tag{6.2}$$

is the *echo-to-background* power ratio (EBR), where background means near-end speech $[v(n)]$ plus ambient noise $[w(n)]$. According to (6.1), the level of convergence (misalignment performance) is completely governed by the EBR. Equation (6.1) is derived by assuming white excitation signal and white ambient noise, and applying some of the results of Chap. 8 — see Sect. 9.4.1 for details.

Regarding question (i) stated above: if we have equally strong talkers at each end, we find that when the echo path has a high attenuation (low EBR) the adaptive filter will be very sensitive to double-talk. On the contrary, a low attenuation means lower sensitivity. Consequently, the EBR is important to consider when setting the parameters of the DTD.

A DTD can be partially characterized by using general detection theory [52], [16], [28]. It is possible to objectively evaluate and compare the performance of different DTDs by means of the probability of correct detection, P_d, and the probability of false alarm, P_f (when double-talk is declared but is not actually present). Moreover, the theory justifies a method to select the threshold T for the detection statistic, ξ.

As far as (ii) is concerned, we can state that a well designed DTD should have a low probability of false alarm when the probability of detection is set to an acceptable level. This level depends on the sensitivity of the adaptive algorithm used in the echo canceler.

Item (iii) can now be addressed in terms of the probability of not detecting (miss) double-talk when it is present, $P_m = 1 - P_d$. The probability of miss is given by,

$$P_m = P\{\xi \geq T | \text{ Double-talk present}\} , \tag{6.3}$$

where $P\{\cdot\}$ denotes "probability of." Assume that we have the convention defined in Sect. 6.2, i.e.,

$$\xi \geq T , \text{ No double-talk}$$
$$\xi < T , \text{ Double-talk} . \tag{6.4}$$

The hypothesis test that the DTD is based on can be formulated as,

$$H_0 : y_e(n) + w(n) , \text{ No double-talk}$$
$$H_1 : y_e(n) + w(n) + v(n) , \text{ Double-talk} . \tag{6.5}$$

Equation (6.2) shows that for constant variance of $y_e(n)$ and $w(n)$, EBR decreases monotonically as the variance of $v(n)$ increases. Therefore from (6.3), (6.4), and (6.5), it follows that P_m increases monotonically with EBR.

To summarize, the performance of both the adaptive filter and the DTD is governed by the EBR. However, while the performance of the adaptive filter degrades with a decreasing EBR, the performance of the DTD improves, so these properties somewhat balance each other. Moreover, the DTD has to be faster, i.e., gather reliable statistics faster, than the adaptive filter. Hence, faster convergence of the AEC results in higher requirements of the DTD. As a consequence, it is important to evaluate the echo canceler and the DTD jointly as a system.

6.2.2 The Geigel DTD

In this section, we give a brief description of a very simple DTD algorithm by A. A. Geigel [37], since it will be used for comparisons in this chapter. The Geigel DTD declares the presence of near-end speech whenever,

$$\xi^{(g)} = \frac{\max\{\, |x(n)|, \dots , |x(n - L_g + 1)|\,\}}{|y(n)|} < T \,, \tag{6.6}$$

where L_g and T are suitably chosen constants. The detection scheme of (6.6) is based on a waveform level comparison between the microphone signal $y(n)$ and the far-end speech $x(n)$, assuming that the near-end speech $v(n)$ at the microphone signal will be typically at the same level, or stronger, than the echo $y_e(n)$. The maximum of the L_g most recent samples of $x(n)$ is taken for the comparison because of the unknown delay in the echo path. The threshold T compensates for the gain of the echo path response h, and is often set to 2 for network echo cancelers because the hybrid (echo path) loss is typically about 6 dB or more. For an AEC, however, it is not easy to set a universal threshold to work reliably in all the various situations because the loss through the acoustic echo path can vary greatly depending on many factors, e.g., placement of microphone relative to the loudspeaker, loudspeaker volume etc. For L_g, one easy choice is to set it the same as the adaptive filter length L since we can assume that the echo path is covered by this length.

6.3 Robust Recursive Least-Squares Algorithms

In Chap. 2, a robust approach to handle double-talk was presented and shown to be very successful in the network echo canceler case, where the combination of outlier-robust adaptive algorithms and a Geigel DTD was studied. For the acoustic case, it is desirable to use a more appropriate DTD, e.g. the NCC DTD, and combine it with a robust adaptive algorithm.

Our objective is to apply the robust recursive least-squares algorithm to multichannel acoustic echo cancellation (see Chap. 5). We will therefore derive a robust multichannel fast recursive least-squares (RLS) algorithm[1]. By *robustness* we mean insensitivity to small deviations of the real distribution from the assumed model distribution [68].

An RLS algorithm can be regarded as a maximum likelihood estimator in which the error distribution has been assumed to be Gaussian. The performance of an algorithm optimized for Gaussian noise could be very poor in the AEC application, because of the unexpected number of large noise values (double-talk disturbances) that are not modeled by the Gaussian law. Thus, when deriving algorithms for the AEC application, the probability density function (PDF) of the noise model should be a long-tailed PDF in order to take the noise bursts due to DTD failures into account. (See [68], [72] for the

---

[1] In general, using the name *least squares* for the robust algorithm is somewhat of a misnomer because we do not minimize the least-squares criterion, but, rather, maximize a likelihood function; the two are synonymous only for Gaussian variates. However, we continue to use the name "least squares" here because of its familiarity.

definition of a long-tailed PDF.) Hence, we are interested in distributional robustness since the shape of the true underlying distribution deviates from the assumed Gaussian model.

In this study, we assume that the PDF of the disturbing noise, $p(z)$, has a tail that is heavier than that of the Gaussian density, $p_G(z)$. By this we mean that $p(z) > p_G(z)$, $\forall |z| \geq |z_0|$. Furthermore, $p(z)$ should be chosen such that the derivative of $\ln[p(z)]$ is bounded as $|z| \to \infty$: although, the difference between $p(z)$ and the Gaussian PDF is small, whether the derivative of $\ln[p(z)]$ is bounded or not makes all the difference between a robust and a non-robust approach.

In the following, we show how to derive a robust RLS algorithm from a PDF, $p(z)$, and how to successfully apply it to the problem of acoustic echo cancellation by combining the resulting algorithm with the fast normalized cross-correlation (FNCC) DTD.

6.3.1 Derivation of the Robust Multichannel RLS

In this section, we derive a robust multichannel RLS algorithm. For this case, we assume that we have P incoming transmission room signals, $x_p(n)$, $p = 1, \ldots, P$. Define the P input excitation vectors as,

$$\mathbf{x}_p(n) = \left[x_p(n)\, x_p(n-1) \cdots x_p(n-L+1) \right]^T , \; p = 1, 2, ..., P ,$$

where the superscript ^{T} denotes transpose of a vector or a matrix. Let the error signal at time n between one arbitrary microphone output $y(n)$ in the receiving room and its estimate be

$$e(n) = y(n) - \sum_{p=1}^{P} \hat{\mathbf{h}}_p^T \mathbf{x}_p(n) , \qquad (6.7)$$

where

$$\hat{\mathbf{h}}_p = \left[\hat{h}_{p,0}\, \hat{h}_{p,1} \cdots \hat{h}_{p,L-1} \right]^T , \; p = 1, 2, ..., P ,$$

are the P modeling filters. In general, we have P microphones in the receiving room as well, i.e., P return channels, and thus P^2 adaptive filters. Here, we just look at one return channel because the derivation is similar for the other $P - 1$ channels. By stacking the P modeling filters and excitation vectors according to,

$$\hat{\mathbf{h}}(n) = \left[\hat{\mathbf{h}}_1^T(n)\, \hat{\mathbf{h}}_2^T(n) \cdots \hat{\mathbf{h}}_P^T(n) \right]^T \qquad (6.8)$$

$$\mathbf{x}(n) = \left[\mathbf{x}_1^T(n)\, \mathbf{x}_2^T(n) \cdots \mathbf{x}_P^T(n) \right]^T , \qquad (6.9)$$

we can derive the multichannel algorithm the same way as for a single-channel algorithm. Thus, we can write

$$e(n) = y(n) - \hat{\mathbf{h}}^T \mathbf{x}(n) . \tag{6.10}$$

The first step in the derivation of a robust algorithm is to choose an optimization criterion. We choose the following function (Chap. 3):

$$J\left(\hat{\mathbf{h}}\right) = \rho \left[\frac{e(n)}{s(n)}\right] , \tag{6.11}$$

where

$$\rho(z) \propto -\ln[p(z)] \tag{6.12}$$

is a convex function (i.e., $\rho''(z) > 0$), $p(z)$ is the PDF of the disturbing noise, and $s(n)$ is a positive scale factor more thoroughly described below.

Newton-type algorithms [75] minimize the optimization criterion (6.11) by the following recursion:

$$\hat{\mathbf{h}}(n) = \hat{\mathbf{h}}(n-1) - \mathbf{R}_{\psi'}^{-1} \nabla J \left[\hat{\mathbf{h}}(n-1)\right] , \tag{6.13}$$

where $\nabla J\left(\hat{\mathbf{h}}\right)$ is the gradient of the optimization criterion w.r.t. $\hat{\mathbf{h}}$, and $\mathbf{R}_{\psi'}$ is (an approximation of) $E\left\{\nabla^2 J\left[\hat{\mathbf{h}}(n-1)\right]\right\}$.

The gradient of $J\left(\hat{\mathbf{h}}\right)$ w.r.t. $\hat{\mathbf{h}}$ is:

$$\nabla J\left(\hat{\mathbf{h}}\right) = -\frac{\mathbf{x}(n)}{s(n)} \psi \left[\frac{e(n)}{s(n)}\right] , \tag{6.14}$$

where,

$$\psi(z) = \rho'(z) = \frac{d}{dz}\rho(z) \propto -\frac{d}{dz}\left\{\ln\left[p(z)\right]\right\} . \tag{6.15}$$

The second derivative (Hessian) of $J\left(\hat{\mathbf{h}}\right)$ is:

$$\nabla^2 J\left(\hat{\mathbf{h}}\right) = \frac{\mathbf{x}(n)\mathbf{x}^T(n)}{s^2(n)} \psi' \left[\frac{e(n)}{s(n)}\right] , \tag{6.16}$$

where,

$$\psi'(z) > 0 , \ \forall z . \tag{6.17}$$

Good choices of $\psi(\cdot)$, and hence of $p(z)$, will be exemplified in the next section.

In this study we choose the following approximation for the expected value of (6.16):

$$\mathbf{R}_{\psi'} = \frac{\mathbf{R}(n)}{s^2(n)} \psi' \left[\frac{e(n)}{s(n)}\right] \approx E\left\{\nabla^2 J\left[\hat{\mathbf{h}}(n-1)\right]\right\} , \tag{6.18}$$

where

$$\mathbf{R}(n) = \sum_{m=1}^{n} \lambda^{n-m} \mathbf{x}(m)\mathbf{x}^T(m)$$
$$= \lambda\mathbf{R}(n-1) + \mathbf{x}(n)\mathbf{x}^T(n) , \tag{6.19}$$

and λ $(0 < \lambda < 1)$ is an exponential forgetting factor. This choice of $\mathbf{R}_{\psi'}$ will allow us to derive a fast version of the robust algorithm.

We can now deduce a robust multichannel RLS algorithm from (6.10), (6.13), (6.14), (6.18), and (6.19) to be,

$$e(n) = y(n) - \hat{\mathbf{h}}^T(n-1)\mathbf{x}(n) , \tag{6.20}$$

$$\hat{\mathbf{h}}(n) = \hat{\mathbf{h}}(n-1) + \frac{s(n)}{\psi'\left[\dfrac{e(n)}{s(n)}\right]}\mathbf{k}(n)\psi\left[\frac{e(n)}{s(n)}\right] , \tag{6.21}$$

$$\mathbf{k}(n) = \mathbf{R}^{-1}(n)\mathbf{x}(n) , \tag{6.22}$$

where $s(n)$ is a scale factor discussed below, $\mathbf{k}(n)$ is the multichannel Kalman gain, and $\mathbf{R}(n)$ is defined in (6.19).

6.3.2 Examples of Robust PDFs

Two examples of PDFs that have the desirable property of bounded first derivative of their log-likelihood functions, and thus are appropriate for modeling the disturbing noise, are:

$$p_1(z) = \frac{1}{2}e^{-\ln[\cosh(\pi z/2)]} , \tag{6.23}$$

$$p_2(z) = \begin{cases} \dfrac{1-\epsilon}{\sqrt{2\pi}}e^{-z^2/2} , & |z| \leq k_0 \\[3mm] \dfrac{1-\epsilon}{\sqrt{2\pi}}e^{-|z|k_0+k_0^2/2} , & |z| > k_0 \end{cases} , \tag{6.24}$$

where the mean and the variance are respectively equal to 0 and 1 in both cases. The second function $p_2(z)$, is known as the PDF of the least informative distribution, [68]. Figure 6.2(a) compares these PDFs to the standard Gaussian density:

$$p_G(z) = \frac{1}{\sqrt{2\pi}}\exp\left\{-z^2/2\right\} . \tag{6.25}$$

It is clearly seen that the tails of $p_1(z)$ and $p_2(z)$ are heavier than that of the Gaussian PDF. Table 6.1 shows the corresponding derivative functions $\psi(\cdot)$ and $\psi'(\cdot)$, and $\psi(\cdot)$ is plotted in Fig. 6.2(b).

Table 6.1 Nonlinear robustness functions and their derivatives. Here, $u(z)$ is the unit-step function, i.e. $u(z) = 1$, $z \geq 0$, $u(z) = 0$, $z < 0$

PDF	$\psi(z)$	$\psi'(z)$
$p_1(z)$	$\pi \tanh(\frac{\pi}{2}z)/2$	$\pi^2 / \left[4\cosh^2(\frac{\pi}{2}z) \right]$
$p_2(z)$	$\max\{-k_0, \min(z, k_0)\}$	$u(z + k_0) - u(z - k_0)$
$p_G(z)$	z	1

6.3.3 Scale Factor Estimation

An important part of the algorithm is the estimation of the scale factor s. Traditionally, the scale is used to make a robust algorithm invariant to the noise level [68]. In our approach, however, it should reflect the minimum mean-square error, be robust to short burst disturbances (double-talk in our application), and track slow changes of the residual error (echo path changes). We have chosen the scale factor estimate as

$$s(n + 1) = \lambda_s s(n) + (1 - \lambda_s) \frac{s(n)}{\psi'\left[\frac{e(n)}{s(n)}\right]} \left| \psi\left[\frac{e(n)}{s(n)}\right] \right| , \qquad (6.26)$$

$$s(0) = \sigma_x .$$

This method of estimating s is justified in [51] and Chaps. 2 and 3.

6.3.4 The Robust Multichannel Fast RLS

A robust multichannel fast RLS (FRLS) algorithm can be derived by using the *a priori* Kalman gain $\mathbf{k}'(n) = \mathbf{R}^{-1}(n-1)\mathbf{x}(n)$. For details of the derivation, see [4]. This *a priori* Kalman gain can be computed recursively with $5P^2L$ multiplications, and the error and adaptation can be computed with $2PL$ multiplications.

"Stabilized" versions of FRLS (with P^2L more multiplications) exist in the literature but they are not significantly more stable than their non-stabilized counterparts for non-stationary signals like speech. Our approach to handle this problem is simply to re-initialize the predictor-based variables when instability is detected with the use of the maximum likelihood variable, $\varphi(n)$, which is inherent in the FRLS computations.

Table 6.2 defines the variables in the robust multichannel FRLS algorithm. Note that in this algorithm, the filter vector $\hat{\mathbf{h}}(n)$ is interleaved, instead of concatenated as in (6.8), and is denoted $\tilde{\mathbf{h}}(n)$. Likewise, the interleaved state vector is denoted as $\tilde{\mathbf{x}}(n)$. This redefinition is done on order to have shift invariance of the correlation matrix in the multichannel case, and thus enables a fast version to be derived. The robust FRLS algorithm with a complexity of $6P^2L + 2PL$ is found in Table 6.3.

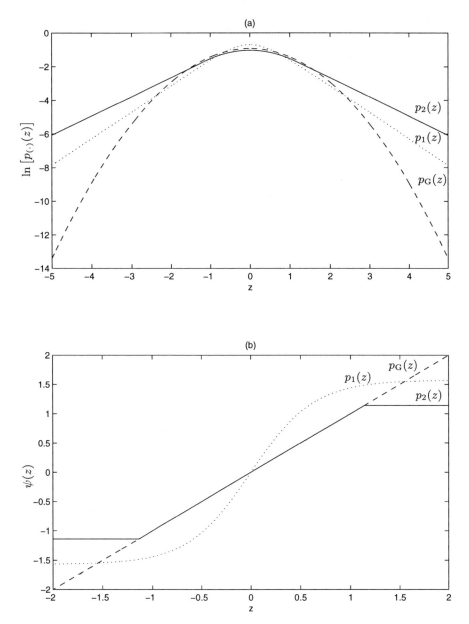

Fig. 6.2. Examples of robust PDFs. (a) PDFs on a logarithmic scale. (b) Corresponding derivative functions, $\psi(z) = -d/dz\{\ln\left[p_{(\cdot)}(z)\right]\}$

6.4 The Normalized Cross-Correlation (NCC) DTD

A decision variable for double-talk detection that has the potential of fulfilling the properties of an "optimum" variable described in Sect. 6.2 was proposed in

Table 6.2 Definitions of variables of the multichannel FRLS algorithm

Variables	Matrix size
$\chi(n) = \begin{bmatrix} x_1(n) \ x_2(n) \ \cdots \ x_P(n) \end{bmatrix}^T$	$(P \times 1)$
$\tilde{x}(n) = \begin{bmatrix} \chi^T(n) \ \ldots \ \chi^T(n-L+1) \end{bmatrix}^T$	$(PL \times 1)$
$\tilde{h}(n) = \begin{bmatrix} \hat{h}_{1,0}(n) \ \hat{h}_{2,0}(n) \ \ldots \ \hat{h}_{P-1,L-1}(n) \ \hat{h}_{P,L-1}(n) \end{bmatrix}^T$	$(PL \times 1)$
$A(n)$, $B(n) =$ Forward and backward prediction matrices	$(PL \times P)$
$E_A(n)$, $E_B(n) =$ Forward and backward prediction error covariance matrices	$(P \times P)$
$e_A(n)$, $e_B(n) =$ Forward and backward prediction error vectors	$(P \times 1)$
$R(n) = \lambda R(n-1) + x(n)x^T(n)$	$(PL \times PL)$
$k(n) = R^{-1}(n)x(n)$, a posteriori Kalman gain vector	$(PL \times 1)$
$k'(n) = R^{-1}(n-1)x(n)$, a priori Kalman gain vector	$(PL \times 1)$
$\alpha(n) = \lambda + x^T(n)R^{-1}(n-1)x(n)$	
$\varphi(n) = \lambda/\alpha(n)$, Maximum likelihood variable	(1×1)
$\kappa \in [1.5, 2.5]$ Stabilization parameter	(1×1)
$\lambda \in]0, 1[$ Forgetting factor	(1×1)

[16], [28]. This variable is formed by normalizing the cross-correlation vector between the excitation vector $x(n)$ and the return signal $y(n)$. Normalization results in a natural threshold level for ξ which is absolutely determined for $v(n) = 0$. In Sect. 6.4.2, we show that this detector can be viewed as a special case of a Neyman-Pearson binary statistical test [111]. The decision variable is given by

$$\xi = \sqrt{r^T(\sigma_y^2 R)^{-1}r} \,, \tag{6.27}$$

where $R = E\{x(n)x^T(n)\}$, $r = E\{x(n)y(n)\}$, and $\sigma_y^2 = E\{y^2(n)\}$. Moreover, we can rewrite ξ as:

$$\xi = \frac{\sqrt{h^T R h}}{\sqrt{h^T R h + \sigma_v^2}} \,, \tag{6.28}$$

by using the identities

$$r = Rh \,, \tag{6.29}$$

and

Table 6.3 The robust multichannel FRLS algorithm

Prediction	Matrix size
$\mathbf{e}_A(n) = \chi(n) - \mathbf{A}^T(n-1)\tilde{\mathbf{x}}(n-1)$	$(P \times 1)$
$\alpha_1(n) = \alpha(n-1) + \mathbf{e}_A^T(n)\mathbf{E}_A^{-1}(n-1)\mathbf{e}_A(n)$	(1×1)
$\begin{bmatrix} \mathbf{t}(n) \\ \mathbf{m}(n) \end{bmatrix} = \begin{bmatrix} \mathbf{0}_{P \times 1} \\ \mathbf{k}'(n-1) \end{bmatrix} +$	$((PL + P) \times 1)$
$\qquad \begin{bmatrix} \mathbf{I}_{P \times P} \\ -\mathbf{A}(n-1) \end{bmatrix} \mathbf{E}_A^{-1}(n-1)\mathbf{e}_A(n)$	
$\mathbf{E}_A(n) = \lambda[\mathbf{E}_A(n-1) + \mathbf{e}_A(n)\mathbf{e}_A^T(n)/\alpha(n-1)]$	$(P \times P)$
$\mathbf{A}(n) = \mathbf{A}(n-1) + \mathbf{k}(n-1)\mathbf{e}_A^T(n)/\alpha(n-1)$	$(PL \times P)$
$\mathbf{e}_{B_1}(n) = \mathbf{E}_B(n-1)\mathbf{m}(n)$	$(P \times 1)$
$\mathbf{e}_{B_2}(n) = \chi(n-L) - \mathbf{B}^T(n-1)\tilde{\mathbf{x}}(n)$	$(P \times 1)$
$\mathbf{e}_B(n) = \kappa\mathbf{e}_{B_2}(n) + (1-\kappa)\mathbf{e}_{B_1}(n)$	$(P \times 1)$
$\mathbf{k}'(n) = \mathbf{t}(n) + \mathbf{B}(n-1)\mathbf{m}(n)$	$(PL \times 1)$
$\alpha(n) = \alpha_1(n) - \mathbf{e}_{B_2}^T(n)\mathbf{m}(n)$	(1×1)
$\mathbf{E}_B(n) = \lambda[\mathbf{E}_B(n-1) + \mathbf{e}_{B_2}(n)\mathbf{e}_{B_2}^T(n)/\alpha(n)]$	$(P \times P)$
$\mathbf{B}(n) = \mathbf{B}(n-1) + \mathbf{k}(n)\mathbf{e}_B^T(n)/\alpha(n)$	$(PL \times P)$
Filtering	
$e(n) = y(n) - \tilde{\mathbf{h}}^T(n-1)\tilde{\mathbf{x}}(n)$	(1×1)
$\tilde{\mathbf{h}}(n) = \tilde{\mathbf{h}}(n-1) + \dfrac{s(n)}{\psi'\left[\dfrac{e(n)}{s(n)}\right]\alpha(n)}\mathbf{k}'(n)\psi\left[\dfrac{e(n)}{s(n)}\right]$	$(PL \times 1)$
$s(n+1) = \lambda_s s(n) + (1-\lambda_s)\dfrac{s(n)}{\psi'\left[\dfrac{e(n)}{s(n)}\right]}\left|\psi\left[\dfrac{e(n)}{s(n)}\right]\right|$	(1×1)

$$\sigma_y^2 = \mathbf{r}^T\mathbf{R}^{-1}\mathbf{r} + \sigma_v^2 \ . \tag{6.30}$$

(Here we assume that the ambient noise $\sigma_w^2 = 0$.) It is easily deduced from (6.28) that for $v(n) = 0$, $\xi = 1$ and for $v(n) \neq 0$, $\xi < 1$. Note also that ξ is independent of the echo path when $v(n) = 0$.

The detection statistic (6.27) is rather complex to compute in real time and therefore unsuitable for practical implementations. However, a fast version of this algorithm can be derived by recursively updating $\mathbf{R}^{-1}\mathbf{r}$ using the Kalman gain $\mathbf{R}^{-1}\mathbf{x}(n)$ [66], which is calculated in the FRLS algorithm. The resulting DTD, called the fast NCC, will be derived next.

Table 6.4 Matrix inversion lemma and other relations

$$\mathbf{R}^{-1}(n) = \lambda^{-1}\mathbf{R}^{-1}(n-1)$$
$$- \lambda^{-1}\frac{\mathbf{R}^{-1}(n-1)\mathbf{x}(n)\mathbf{x}^T(n)\mathbf{R}^{-1}(n-1)}{\alpha(n)}$$
$$\mathbf{k}(n) = \mathbf{R}^{-1}(n)\mathbf{x}(n)$$
$$= \frac{1}{\alpha(n)}\mathbf{R}^{-1}(n-1)\mathbf{x}(n) , \quad \text{a posteriori Kalman gain}$$
$$\mathbf{k}'(n) = \mathbf{R}^{-1}(n-1)\mathbf{x}(n) , \quad \text{a priori Kalman gain}$$
$$\alpha(n)\mathbf{x}^T(n)\mathbf{k}(n) = \mathbf{x}^T(n)\mathbf{R}^{-1}(n-1)\mathbf{x}(n)$$
$$\alpha(n)\mathbf{r}^T(n-1)\mathbf{k}(n) = \mathbf{r}^T(n-1)\mathbf{R}^{-1}(n-1)\mathbf{x}(n)$$
$$= \hat{\mathbf{h}}_b^T(n-1)\mathbf{x}(n) = \hat{y}(n) , \quad \text{estimated echo}$$

6.4.1 Fast NCC (FNCC)

Estimated quantities of the cross-correlation and the near-end signal power are introduced for the derivation of the FNCC DTD. Equation (6.27) can be rewritten as

$$\xi^2(n) = \frac{\mathbf{r}^T(n)\mathbf{R}^{-1}(n)\mathbf{r}(n)}{\sigma_y^2(n)} = \frac{\eta^2(n)}{\sigma_y^2(n)} , \tag{6.31}$$

where we squared the statistic for simplicity. The correlation variables are estimated as,

$$\mathbf{r}(n) = \lambda\mathbf{r}(n-1) + \mathbf{x}(n)y(n) , \tag{6.32}$$
$$\mathbf{R}(n) = \lambda\mathbf{R}(n-1) + \mathbf{x}(n)\mathbf{x}^T(n) , \tag{6.33}$$
$$\sigma_y^2(n) = \lambda\sigma_y^2(n-1) + y^2(n) . \tag{6.34}$$

Table 6.4 reviews some useful relations that are used in the following, where $\alpha(n)$ is defined in Table 6.2. Note that we need to distinguish between the "background" echo path estimate calculated in the FNCC DTD, from now on denoted $\mathbf{h}_b(n)$, and the estimate calculated in the echo canceler, $\mathbf{h}(n)$.

Let's now look at the statistic $\eta^2(n)$,

$$\eta^2(n) = \left[\lambda \mathbf{r}^T(n-1) + y(n)\mathbf{x}^T(n)\right] \mathbf{R}^{-1}(n) \left[\lambda \mathbf{r}(n-1) + \mathbf{x}(n)y(n)\right]$$

$$= \lambda^2 \mathbf{r}^T(n-1)\mathbf{R}^{-1}(n)\mathbf{r}(n-1) + \lambda \mathbf{r}^T(n-1)\mathbf{R}^{-1}(n)\mathbf{x}(n)y(n)$$

$$\quad + \lambda y(n)\mathbf{x}^T(n)\mathbf{R}^{-1}(n)\mathbf{r}(n-1) + y^2(n)\mathbf{x}^T(n)\mathbf{R}^{-1}(n)\mathbf{x}(n)$$

$$= \lambda^2 \mathbf{r}^T(n-1)\left[\lambda^{-1}\mathbf{R}^{-1}(n-1) - \lambda^{-1}\alpha(n)\mathbf{k}(n)\mathbf{k}^T(n)\right]\mathbf{r}(n-1)$$

$$\quad + 2\lambda y(n)\mathbf{k}^T(n)\mathbf{r}(n-1) + y^2(n)\mathbf{x}^T(n)\mathbf{k}(n)$$

$$= \lambda \mathbf{r}^T(n-1)\mathbf{R}^{-1}(n-1)\mathbf{r}(n-1) - \lambda\alpha(n)\left[\mathbf{k}^T(n)\mathbf{r}(n-1)\right]^2$$

$$\quad + 2\lambda y(n)\mathbf{k}^T(n)\mathbf{r}(n-1) + y^2(n)(1 - \frac{\lambda}{\alpha(n)})$$

$$= \lambda \eta^2(n-1) - \frac{\lambda}{\alpha(n)}\hat{y}^2(n) + 2\frac{\lambda}{\alpha(n)}y(n)\hat{y}(n) + (1 - \frac{\lambda}{\alpha(n)})y^2(n)$$

$$= \lambda \eta^2(n-1) - \varphi(n)\hat{y}^2(n) + 2\varphi(n)\hat{y}(n)y(n) + [1 - \varphi(n)]y^2(n)$$

$$= \lambda \eta^2(n-1) + y^2(n) - \varphi(n)e^2(n) , \tag{6.35}$$

where the likelihood variable $\varphi(n) = \lambda/\alpha(n)$ and $e(n)$ is the residual error, $e(n) = y(n) - \hat{y}(n)$. Thus, we find that the quantities needed to form the test statistic of the FNCC DTD are given by the simple first-order recursions in (6.34) and (6.35). Table 6.5 gives the essential calculations for both DTD and echo canceler, where it is assumed that the Kalman gain has been calculated by the FRLS algorithm in Table 6.3, and $\hat{\tilde{\mathbf{h}}}_{\mathrm{b}}(n)$ is an interleaved version of $\hat{\mathbf{h}}_{\mathrm{b}}(n)$ as in Sect. 6.3.4. (Also, recall the interleaved definitions $\tilde{\mathbf{h}}$ and $\tilde{\mathbf{x}}$ used there.)

6.4.2 The NCC DTD as a Neyman-Pearson Detector

Assume that we have the following binary hypothesis test:

$$H_0 : y(n) = \mathbf{h}^T\mathbf{x}(n) \tag{6.36}$$

$$H_1 : y(n) = \mathbf{h}^T\mathbf{x}(n) + v(n) \tag{6.37}$$

and let[2]

$$\mathbf{y}(n) = [y(n)\ y(n-1)\ \cdots\ y(n-N+1)]^T . \tag{6.38}$$

The Neyman-Pearson detection statistic (likelihood function) is then given by,

$$L\left[\mathbf{y}(n)\right] = \frac{p\left[\mathbf{y}(n)|H_1\right]}{p\left[\mathbf{y}(n)|H_0\right]} \underset{<}{\overset{>}{\gtrless}} \gamma'' , \tag{6.39}$$

where γ'' is a general detection threshold. Assuming white Gaussian signals[3], we find the following log-likelihood function:

---

[2] For simplicity the ambient noise is assumed to be zero in this section.

[3] This is a very crude approximation of the real situation which is neither Gaussian nor white.

Table 6.5 The FNCC double-talk detector and generic robust echo canceler

Double-talk detector	Multiplications
$\sigma_y^2(n) = \lambda\sigma_y^2(n-1) + y^2(n)$	2
$e_b(n) = y(n) - \tilde{\mathbf{h}}_b^T(n-1)\tilde{\mathbf{x}}(n)$	PL
$\eta^2(n) = \lambda\eta^2(n-1) + y^2(n) - \varphi(n)e_b^2(n)$	3

$\eta(n)/\sigma_y(n) < T$, $\Rightarrow$ double-talk, $\mu = 0$

$\eta(n)/\sigma_y(n) \geq T$, $\Rightarrow$ no double-talk, $\mu = 1$

$$\tilde{\mathbf{h}}_b(n) = \tilde{\mathbf{h}}_b(n-1) + \mathbf{k}'(n)\frac{e_b(n)}{\varphi(n)} \qquad PL$$

Robust echo cancellation

$$e(n) = y(n) - \tilde{\mathbf{h}}^T(n-1)\tilde{\mathbf{x}}(n) \qquad PL$$

$$\tilde{\mathbf{h}}(n) = \tilde{\mathbf{h}}(n-1) + \mu\frac{s(n)}{\psi'\left[\frac{e(n)}{s(n)}\right]\varphi(n)}\mathbf{k}'(n)\psi\left[\frac{e(n)}{s(n)}\right] \qquad PL$$

$$s(n+1) = \lambda_s s(n) + (1-\lambda_s)\frac{s(n)}{\psi'\left[\frac{e(n)}{s(n)}\right]}\left|\psi\left[\frac{e(n)}{s(n)}\right]\right|$$

$$l\left[\mathbf{y}(n)\right] = \ln L\left[\mathbf{y}(n)\right]$$

$$= \underbrace{\frac{N}{2}\ln\left(\frac{\sigma_{h^T x}^2}{\sigma_{h^T x}^2 + \sigma_v^2}\right) + \frac{1}{2}\frac{\sigma_{h^T x}^2}{\sigma_{h^T x}^2(\sigma_{h^T x}^2 + \sigma_v^2)}}_{\text{Independent of data}}\mathbf{y}^T(n)\mathbf{y}(n) \gtrless \gamma' ,$$

which is equivalent to

$$l'\left[\mathbf{y}(n)\right] = \mathbf{y}^T(n)\mathbf{y}(n) \gtrless \gamma , \tag{6.40}$$

since only the term $\mathbf{y}^T(n)\mathbf{y}(n)$ is data dependent. To make the statistic independent of $\mathbf{h}$ we can normalize it with $\hat{\mathbf{y}}^T(n)\hat{\mathbf{y}}(n)$ where

$$\hat{\mathbf{y}}(n) = [\hat{y}(n)\ \hat{y}(n-1)\ \cdots\ \hat{y}(n-N+1)]^T \tag{6.41}$$

and

$$\hat{y}(n) = \hat{\mathbf{h}}^T\mathbf{x}(n) . \tag{6.42}$$

Furthermore, we invert the result to be consistent with our threshold convention, which gives:

$$\xi^2 = \frac{\hat{\mathbf{y}}^T(n)\hat{\mathbf{y}}(n)}{l'\left[\mathbf{y}(n)\right]} = \frac{\hat{\mathbf{y}}^T(n)\hat{\mathbf{y}}(n)}{\mathbf{y}^T(n)\mathbf{y}(n)} . \tag{6.43}$$

We note from (6.38) that $E\left[\mathbf{y}^T(n)\mathbf{y}(n)\right] = N\sigma_y^2$. Furthermore, assuming that $\hat{\mathbf{h}}(n)$ has converged to a close approximation of $\mathbf{h}$, and using (6.29) with (6.41) and (6.42), we have, ideally, $E\left[\hat{\mathbf{y}}^T(n)\hat{\mathbf{y}}(n)\right] \approx N\mathbf{r}^T\mathbf{R}^{-1}\mathbf{r}$, where $\mathbf{R}$ and $\mathbf{r}$ are the true covariance matrix and crosscorrelation vector respectively. Therefore, under these ideal conditions, (6.43) is roughly equivalent to (6.31). This shows that the NCC detector is an exponential window implementation of the Neyman-Pearson binary hypothesis test.

6.4.3 The Link Between FNCC and the Two-Path DTD Scheme

One common way to handle double-talk is to use the two-path structure of [104]. In this case there is a background adaptive filter modeling the echo path and a fixed foreground filter which cancels the echo. Whenever the background filter is performing well according to some criterion, it is copied to the foreground filter. An interesting interpretation of the echo canceler and FNCC DTD shown in Table 6.5 is that it naturally has a two-path structure: the DTD represents the background adaptive filter, and the AEC a foreground filter. Hence, the derivation of the FNCC naturally leads to a foreground/background structure. What distinguishes the FNCC DTD from the two-path structure of [104] is that the foreground filter is also adaptive and the background weights, $\mathbf{h}_b$, are not copied to the foreground filter.

6.5 Practical Modifications of the FNCC DTD and FRLS Echo Canceler

Sections 6.3.4 and 6.4.1 give the theoretical derivations of the robust FRLS and the FNCC detector. In a realistic situation with non-stationary input (speech), some of the assumptions used in these derivations are violated. We have therefore slightly modified the algorithms in order to handle these problems better.

Table 6.6 presents a modified version of the FNCC DTD and robust FRLS used for the simulations in this section. To be able to freely choose the response of the FNCC detector faster than that of the FRLS algorithm, we introduce a new forgetting factor, λ_1, in (6.46) and (6.47). A preferable choice is $\lambda_1 = 1 - 2/L$. Theoretically, (6.31) should always be between 0 and 1. However, in the practical algorithm where (6.35) is used, this property can be violated when the background filter is seriously perturbed, e.g., by double-talk or by an echo path change. Equations (6.45)–(6.49) modify the detector such that $\eta^2(n)$ and $\sigma_y^2(n)$ are not updated when $y^2(n) - \varphi(n)e^2(n)$ in (6.35) is negative.

Equations (6.44) and (6.50) are new foreground and background energy estimators. If the background filter is performing better than the foreground, and at the same time the nonlinear robustness function is in its saturated region, we know that an echo path change has occurred; in that case, convergence is increased by reducing $\psi'(\cdot)$. This control is implemented in (6.51)–(6.53).

Finally, robustness is maintained, in spite of detection misses, over longer periods of double-talk by keeping the scale factor small, (6.54)–(6.56).

6.6 Evaluation and Simulations

The adaptive algorithm for the acoustic echo canceler was derived for the multichannel case in this chapter. In these simulations, however, we prefer to do most of the evaluations using the single-channel version since these simulations are less complex to perform. However, for the final simulation, we do include the performance of the robust two-channel algorithm for a double-talk situation. More generally, all results are directly applicable to the multichannel case.

6.6.1 Receiver Operating Characteristic (ROC)

When evaluating a detector, it is customary to show the probability of detection P_d or, equivalently, the probability of miss $P_m = 1 - P_d$, versus the probability of false alarm, P_f. By "detection," we mean a correct decision is made when double-talk is present. A "miss" is defined as an incorrect decision when double-talk is present, and a "false alarm" is the decision to declare double-talk when it is not present.

The curve showing the probability of detection (or miss) versus the probability of false alarm is known as the receiver operating characteristic (ROC) of a detector. This curve has been widely used for characterizing detectors in communication systems and radar.

We adopt the ROC technique for characterizing double-talk detectors, as previously suggested in [52], [28]. In our presentations, we prefer to show the probability of miss versus the probability of false alarm since the miss and false alarm probabilities are usually in the same numerical range. One problem that has to be kept in mind, though, is that speech has a time-varying power level. This may give less consistent estimates of P_m and P_f than in the case of a stationary signal. Nevertheless, for comparative studies of the DTDs, the estimated ROC is still a useful performance metric.

Estimates of P_m and P_f using a speech database were obtained according to the procedure in [28]. The data we use contain sentences from three male and two female talkers. Furthermore, all sentences are also normalized to have the same average power level. Simulation details are as follows:

Echo path. The echo path used is a measured acoustic response between the left loudspeaker and a standard cardioid microphone positioned on top of a workstation. The original impulse response has a length of 256 ms, consisting of 4096 coefficients at a 16 kHz sampling rate. However, it was subsequently decimated to an 8 kHz sampling rate, resulting in 2048 coefficients (see Fig. 6.3). It is also normalized so that $\sigma_{y_e}^2 = \sigma_x^2$ for the actual speech data.

Table 6.6 The FNCC double-talk detector and FRLS echo canceler with modifications used in all simulations

Double-talk detector	**Multipl.**	
$e_b(n) = y(n) - \tilde{\mathbf{h}}_b^T(n-1)\tilde{\mathbf{x}}(n)$	PL	
$\sigma_{e_b}^2(n) = \lambda_1 \sigma_{e_b}^2(n-1) + e_b^2(n)$	2	(6.44)
if $y^2(n) \geq \varphi(n)e_b^2(n)$		(6.45)
$\quad \sigma_y^2(n) = \lambda_1 \sigma_y^2(n-1) + y^2(n)$	2	(6.46)
$\quad \eta^2(n) = \lambda_1 \eta^2(n-1) + y^2(n) - \varphi(n)e_b^2(n)$	3	(6.47)
else		
$\quad \sigma_y^2(n) = \sigma_y^2(n-1)$		(6.48)
$\quad \eta^2(n) = \eta^2(n-1)$		(6.49)
end		
$\eta(n)/\sigma_y(n) < T ,\ \Rightarrow$ double-talk, $\mu = 0$		
$\eta(n)/\sigma_y(n) \geq T ,\ \Rightarrow$ no double-talk, $\mu = 1$		
$\tilde{\mathbf{h}}_b(n) = \tilde{\mathbf{h}}_b(n-1) + \mathbf{k}'(n)\dfrac{e_b(n)}{\varphi(n)}$	PL	
Robust echo cancellation		
$e(n) = y(n) - \tilde{\mathbf{h}}^T(n-1)\tilde{\mathbf{x}}(n)$	PL	
$\sigma_e^2(n) = \lambda_1 \sigma_e^2(n-1) + e^2(n)$	2	(6.50)
if $\sigma_{e_b}^2(n) < \sigma_e^2(n)/\sqrt{2}$ & $\left\|\psi\left[\dfrac{e(n)}{s(n)}\right]\right\| = k_0$		(6.51)
$\quad \psi_f'\left[\dfrac{e(n)}{s(n)}\right] = 0.5$		(6.52)
else		
$\quad \psi_f'\left[\dfrac{e(n)}{s(n)}\right] = 1$		(6.53)
end		
$\tilde{\mathbf{h}}(n) = \tilde{\mathbf{h}}(n-1) + \mu \dfrac{s(n)}{\psi_f'\left[\dfrac{e(n)}{s(n)}\right]\varphi(n)}\mathbf{k}'(n)\psi\left[\dfrac{e(n)}{s(n)}\right]$	PL	
if (No double-talk)		(6.54)
$\quad s(n+1) = \lambda_s s(n) + (1-\lambda_s)\dfrac{s(n)}{\psi_f'\left[\dfrac{e(n)}{s(n)}\right]}\left\|\psi\left[\dfrac{e(n)}{s(n)}\right]\right\|$	3	(6.55)
else		
$\quad s(n+1) = \lambda_s s(n) + (1-\lambda_s)s_{\min}$	2	(6.56)
end		

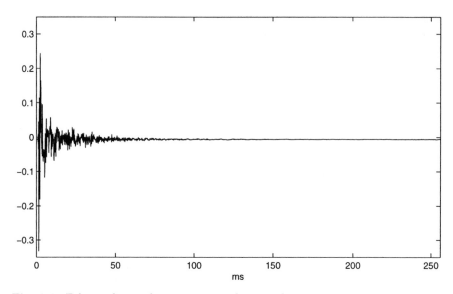

Fig. 6.3. Echo path impulse response used in simulation

Probability of false alarm. When estimating the probability of false alarm we use sentences from five talkers as far-end speech. These consist of speech sequences of 21.8 seconds at an 8 kHz sampling rate. The echo-to-ambient-noise ratio, ENR $= \sigma_{y_e}^2/\sigma_w^2$, is set to 1000 (30 dB).

Probability of miss. The probability of a miss is estimated using 5 seconds of far-end speech from one male talker. As near-end speech, 8 sentences are used, each about 2 seconds long. In our case we investigate the performance when the average echo-to-background ratio EBR $= \sigma_{y_e}^2/(\sigma_v^2 + \sigma_w^2) = 1$ (0 dB), since it is natural to assume equally strong talkers.

The thresholds of the Geigel and FNCC detectors are chosen such that their probability of false alarm range from approximately 0 to about 0.25. For these thresholds, we then estimate the corresponding probability of detection (see [28] for details). Results from these simulations are shown in Fig. 6.4. These results are consistent with those found in [28] and, by the use of the ROC, it is possible to set thresholds such that the DTDs are compared fairly.

6.6.2 AEC/DTD Performance During Double-Talk

Our purpose here is to show that the FNCC DTD is very suitable for the acoustic case. For these simulations, we use a subset of the speech data described in Sect. 6.6.1. Data and algorithm settings are:

Far-end: Speech from one male talker repeated twice giving a total of 10 seconds of data. The standard deviation of the signal is $\sigma_x = 1900$.

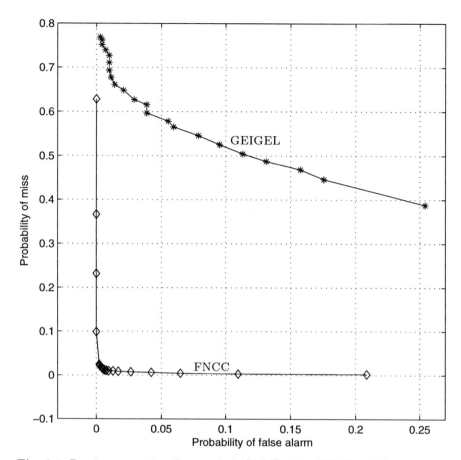

Fig. 6.4. Receiver operating characteristic (ROC) of the FNCC and Geigel DTDs. EBR = 0 dB and ENR = 30 dB

Near-end: Speech from another male talker.
Levels: EBR = 1 (0 dB), ENR = 1000 (30 dB).
Echo path: The same as in the previous section.
FRLS parameters: $L = 1024$ (128 ms), $\lambda = 1 - 1/3L$, $\kappa = 0$, $\hat{\mathbf{h}}_0 = \mathbf{0}$.
FNCC parameters: $\lambda_1 = 1 - 2/L$, $T = 0.85 \Rightarrow P_{\mathrm{m}} = 0.63$, $P_{\mathrm{f}} \approx 0$ (too small to be reliably measured — see Fig. 6.4).
Geigel DTD parameter: $T = 1.20 \Rightarrow P_{\mathrm{m}} = 0.63$, $P_{\mathrm{f}} \approx 0.03$.
Robust settings: $\psi \left[\dfrac{e(n)}{s(n)} \right] = \max \left\{ -k_0, \min \left[\dfrac{e(n)}{s(n)}, k_0 \right] \right\}$, $\lambda_{\mathrm{s}} = 0.9975$, $k_0 = 1.5$, $\beta \approx 0.60655$, $s_0 = 1000$.

The choice of thresholds for the detectors is based on practical aspects. In general, lower P_{m} is achieved at the cost of higher P_{f}, which necessitates a trade-off in performance depending on the penalty or cost function of a false alarm and of a miss. For conventional (non-robust) AEC's, the penalty of false alarm is small because it simply halts the filter adaptation for t_{hold} seconds, incurring only a P_{f} percentage delay of the convergence time. Also, conventionally, low P_{m} is important to prevent divergence due to double-talk. However, for robust AEC's the situation is reversed: with robustness, a low P_{m} is no longer paramount, whereas a low P_{f} becomes more important because of the slower recovery from echo path changes. Thus, for the robust AEC, it is preferable to have a low probability of false alarm and a moderate probability of detection rather than a high probability of detection and higher false alarm rate. Normally one would compare detectors set to have the same probability of false alarm. However, for a reasonable detection rate (0.35-0.7) the FNCC DTD has a false alarm rate that is virtually zero in our conditions. Forcing the Geigel DTD to have such a low false alarm rate will result in a miss rate close to 1! We therefore choose thresholds for the DTDs such that they have the same probability of miss. This of course has the effect that the Geigel detector exhibits a higher false alarm rate.

The performance is measured by means of the normalized mean-square error (MSE), which we estimate according to

$$\text{MSE}(n) = \frac{\text{LPF}([y_{\text{e}}(n) - \hat{y}(n)]^2)}{\text{LPF}([y_{\text{e}}(n)]^2)} , \tag{6.57}$$

where LPF denotes a first-order lowpass filter with a pole at 0.999 (time constant $= 1/16\,\text{s}$ at $16\,\text{kHz}$ sampling rate).

Figure 6.5 shows far-end speech, double-talk, and mean-square error performance of the FRLS and robust FRLS algorithms when either the FNCC or the Geigel detector is used as a DTD. The time intervals for which the DTDs have indicated double-talk are also shown below the MSE-curves. From these results we draw the following conclusions: Looking at the MSE-performance of the *non-robust* FRLS algorithm, both detectors miss the very first parts of the double-talk and therefore the (non-robust) FRLS AEC algorithm diverges. During the double-talk period, detection misses of the Geigel DTD result in further performance loss for the canceler. No divergence is seen for the *robust* FRLS, regardless of DTD, since the robustness handles the situation when there are detection misses.

We note that, for this case of a stationary echo path, there is not much difference between the FNCC and Geigel DTD's. However, significant differences will emerge for the case of changing echo paths, as studied in the following section.

6.6.3 AEC/DTD Sensitivity to a Time-Varying Echo Path

For the purpose of showing how the DTD/AEC system handles echo path changes, we have chosen to study four cases. The same speech data and algorithm settings as in the previous section are used, but no double-talk is

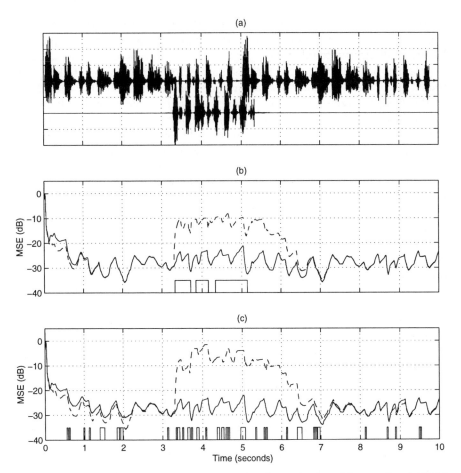

Fig. 6.5. Double-talk situation of robust FRLS (Solid line) and (non-robust) FRLS (Dashed line). (a) Far-end speech (upper), near-end speech (lower). (b) Mean-square error with the FNCC DTD. (c) Mean-square error with the Geigel DTD. The rectangles in the bottom of (b) and (c) indicate where double-talk has been detected

introduced. In each simulation, the echo path undergoes one of four changes after the FRLS algorithm has been adapting for 2 seconds:

a)   Time delay: $h(n) \rightarrow h(n-5)$.
b)   Time advance: $h(n) \rightarrow h(n+14)$.
c)   6 dB increase of echo path gain: $h(n) \rightarrow 2h(n)$.
d)   6 dB decrease of echo path gain: $h(n) \rightarrow 0.5h(n)$.

In reality, there are infinite possibilities for echo path variations. However, these four cases were chosen to represent the basic variations that may occur.

A major concern, when an echo path change occurs, is that the false alarm rate of the DTD may increase. That would undoubtedly reduce the convergence rate. We note that, in general, the robust AEC is worse than the non-robust AEC in adapting to echo path changes; that is the price paid for better immunity to double-talk.

Figures 6.6 and 6.7 show mean-square error performance of the (non-robust) FRLS and robust FRLS algorithms combined with either the FNCC or the Geigel detector as DTD. From these curves we draw the following conclusions: The FNCC DTD (Fig. 6.6) does not have a significant increase of false alarms due to echo path changes; only for the time advance change, Fig. 6.6(b), is a false alarm event visible. (Actually the event consists of several false alarms, which cannot be resolved on this time scale.) With use of the foreground/background structure in the FNCC/robust FRLS combination, we detect the echo path change and improve the convergence rate, through $\psi'\{\cdot\}$, of the foreground robust FRLS algorithm such that it becomes almost as fast as the (non-robust) FRLS algorithm.

The Geigel DTD has about the same false alarm rate after time delay/advance echo path changes [Fig. 6.7(a), (b)]. The convergence rate of both robust and non-robust FRLS is already affected by these false alarms [compare with Fig. 6.6(a), (b)]. When the echo path gain increases [Fig. 6.7(c)], the false alarm rate becomes so high that the adaptation of the robust FRLS is halted and the convergence time of the (non-robust) FRLS algorithm is tripled. For a decrease in echo path gain, there is of course not as much degradation of the Geigel DTD performance as in the other cases. This is due to the fact that our assumption of echo path attenuation in the Geigel DTD is fulfilled (with margin) in this case.

6.6.4 Two-Channel AEC/DTD Performance During Double-Talk

For this stereo simulation, we use real-life speech data described in [41]. These recordings were made in an actual conference room facility [21]. The microphone and loudspeaker setup in the near-end room are shown in Fig. 6.8, and the estimated acoustic responses and magnitude functions are shown, respectively, in Figs. 6.9 and 6.10. In order to have a unique solution for the echo canceler, the far-end signals have been processed with a nonlinearity according to [19] (see Chap. 5). The original data sampled at 16 kHz was down-sampled by a factor of 2 in order to have a sampling rate of 8 kHz. Furthermore, we added near-end speech to the microphone signals for the purpose of showing the double-talk behavior of the robust two-channel FRLS/FNCC DTD.

The basic speech and algorithm settings are:

Far-end: Stereo recordings with the same male talker as in the previous section. The distortion parameter of the nonlinearity is 0.5 [19], Chap. 5. That is, the far-end channels are pre-processed according to:

$$x_{(\cdot)} = x_{(\cdot)} + \frac{0.5}{2}\left[x_{(\cdot)} \pm |x_{(\cdot)}|\right], \tag{6.58}$$

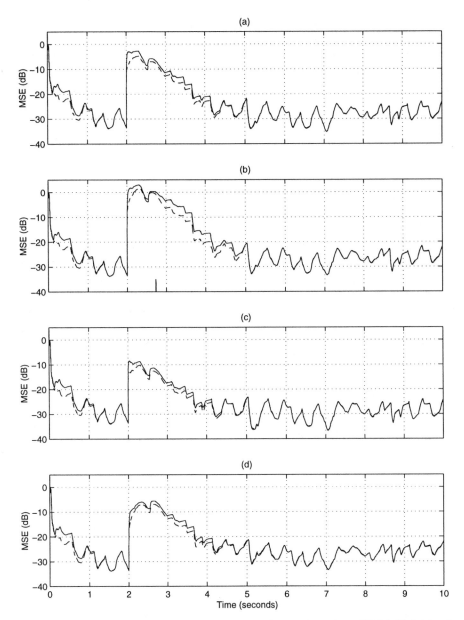

Fig. 6.6. Mean-square error during echo path change for robust FRLS (Solid line) and (non-robust) FRLS (Dashed line), both with the FNCC DTD. The echo path changes at 2 seconds: (a) Echo path time delay of 5 samples. (b) Echo path time advance of 14 samples. (c) 6 dB increase of echo path gain. (d) 6 dB decrease of echo path gain

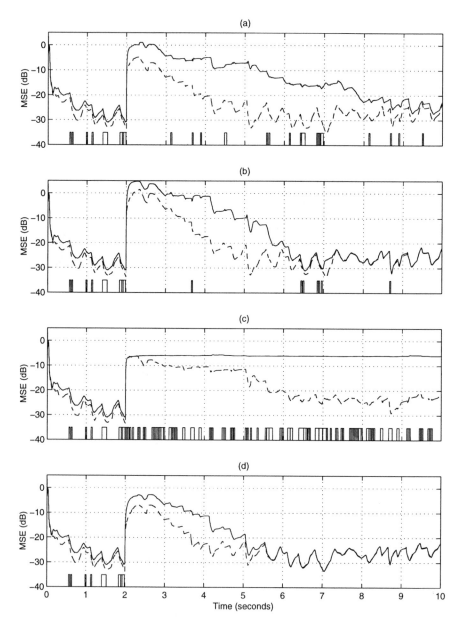

Fig. 6.7. Mean-square error of the robust FRLS (Solid line) and (non-robust) FRLS (Dashed line), both with the Geigel DTD. Other conditions as in Fig. 6.6

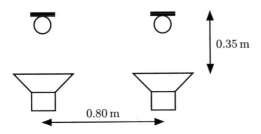

Fig. 6.8. Near-end room microphone and loudspeaker setup

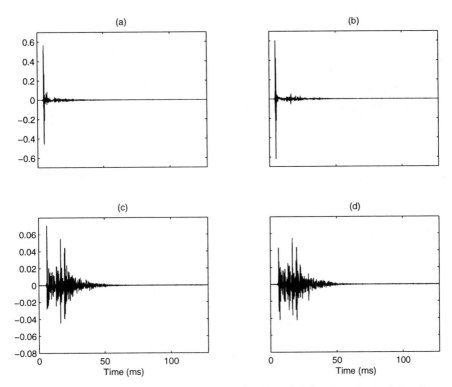

Fig. 6.9. Near-end room impulse responses. (a) From left loudspeaker to left microphone. (b) From right loudspeaker to right microphone. (c) From left loudspeaker to right microphone. (d) From right loudspeaker to left microphone

where $(.)$ denotes either $(left)$ or $(right)$ channel. The positive half-wave is added to the left channel and the negative to the right.

Near-end: Speech from another male talker.

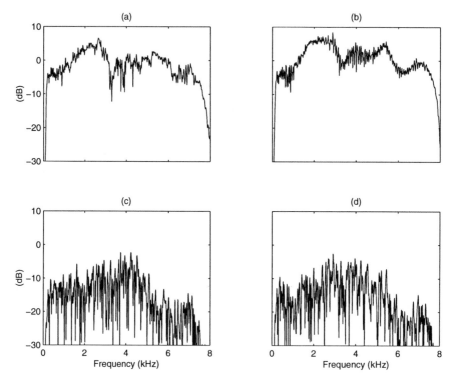

Fig. 6.10. Frequency response magnitude of the near-end room. Other conditions as in Fig. 6.9

Levels: EBR ≈ 1 (0 dB), ENR ≈ 6000 (38 dB).
FRLS parameters: $L = 1024$ (128 ms), $\lambda = 1 - 1/(3 \cdot 2L)$, $\kappa = 1$, $\hat{\mathbf{h}}_0 = \mathbf{0}$.
FNCC parameters: $\lambda_1 = 1 - 1/L$, $T = 0.90 \Rightarrow P_m = 0.37$, $P_f \approx 0$
(Fig. 6.4).

Figure 6.11 shows the mean-square error performance during double-talk. We find that the two-channel algorithm behaves as well as the single-channel algorithm presented in the previous section.

6.7 Conclusions

In this chapter, we have combined methods from robust statistics with recursive least-squares filtering and double-talk detection in order to solve problems that arise in acoustic echo cancellation. An acoustic echo cancellation system comprising a robust multichannel fast RLS adaptive algorithm and a fast normalized cross correlation double-talk detector has been presented. Major results of the FNCC detector are:

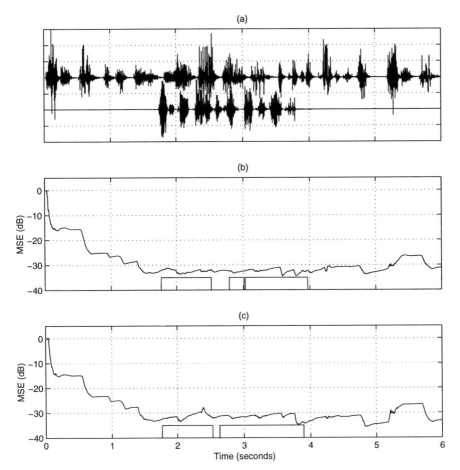

Fig. 6.11. Double-talk situation. Robust two-channel FRLS with FNCC DTD. (a) Left channel far-end speech (upper), near-end speech (lower). (b) Mean-square error of left channel. (c) Mean-square error of right channel

- Much higher accuracy compared to the Geigel DTD. That is, lower false alarm rate at a fixed probability of detection.
- Not dependent on any assumptions about the echo path, e.g., the echo path gain. This is a very important property since acoustic echo paths vary quite a bit.
- Low added complexity when implemented together with the FRLS algorithm.
- The required nonlinear function $[\psi(\cdot)]$ in the robust FRLS results in some minor performance loss (slower convergence rate) compared to the non-robust FRLS. However, we can easily compensate for this loss by combining the FNCC decision with some simple statistics (see Table 6.6).

Furthermore, theoretical analysis and simulations have shown that:

- The derivation of the FNCC naturally leads to a foreground/background dualism that is functionally similar to the well-known two-path technique.
- Requiring a higher probability of detection (double-talk), i.e., increasing the threshold, increases the false alarm rate after echo path changes.
- Increasing λ in FRLS will not result in as low divergence as that of the robust FRLS during double-talk, provided that both algorithms are constrained to have the same convergence rate.

The combination of robust statistics with an optimal[4] adaptive algorithm (FRLS), and the FNCC DTD results in a highly effective system where theoretical and practical aspects complement each other, e.g., robustness makes it possible to lower the probability of detection of the DTD. This in turn results in better performance (lower false alarm rate) of the DTD during echo path changes and thus faster convergence of the echo canceler.

The FNCC detector uses intermediate calculations of the FRLS algorithm. Another way of implementing this detector is to base the detection statistic on a projection of the near-end data onto a subspace spanned by the far-end signal. An algorithm that relies on this principle is the affine projection algorithm [106] (see Chap. 5). In this case, we can use the same techniques as the fast affine projection algorithm [55], in order to have a fast version of the DTD. It is also noted that for a projection order equal to one, we get the normalized least mean-square (NLMS) algorithm which would result in a very low complexity system of DTD and AEC.

So far, we have confined our study to the fullband case of adaptive filtering. The ideas presented could of course be generalized to a subband or frequency-domain scheme. Subband implementations are described in Chap. 7, and in Chaps. 8 and 9 the frequency-domain approach to multichannel adaptive filtering and double-talk detection is pursued.

---

[4] In the sense that it is a maximum likelihood algorithm if the disturbance is distributed according to our assumptions.

7. Some Practical Aspects of Stereo Teleconferencing System Implementation

7.1 Introduction

When people with normal hearing converse in a room where many people are speaking simultaneously, their binaural hearing enables them to focus in on particular talkers according to the directions from which those talkers' voices are coming. They can do this even when the signal-to-background-noise ratio is very low; noise, in this case being the voices of those ignored. This phenomenon of human audio perception is aptly called the *cocktail party effect*. In monophonic teleconferencing systems, this aid to audio communication is lost across the connection. A listener on one side hears all of the talkers of the far side coming from the same direction – the direction from a single local loudspeaker. So, when people on the far side talk simultaneously, it is impossible for the local listener to spatially separate their voices as he or she normally would. A stereo connection would solve this problem because with stereo the local listener (when located in the "sweet spot" – that area where the stereo effect is most clearly perceived) hears a reconstruction of the left-right positioning of the sound from the far side. Until very recently though, teleconferencing systems have been limited to monophonic connections because stereo acoustic echo cancellation was problematic [121]. However, with the advent of new techniques (see Chap. 5) such systems are now quite realizable.

Very often teleconferences are multi-point rather than simple point-to-point connections. Such multi-point connections are accomplished with so-called *conference bridges*. There are several ways that stereo audio can be exploited in this situation. Consider an N-point conference bridge. Arbitrarily select one of the endpoints as the near endpoint, the others being far endpoints. One way for the bridge to operate would be to take the monophonic audio signals from the $N-1$ far endpoints and synthesize a stereo signal from each (placing it in a particular lateral position), sum them, and send the result to the near endpoint [18]. The lateral audio positioning of each of the far endpoints could be derived from the lateral video positioning of those endpoints on the near-end video display. Then, if the near-end user should move the video position of a far endpoint, the audio would "follow" it. The latter location information from the near endpoint can be sent to the conference bridge over a low-bandwidth control channel.

Multi-point teleconferences can be further enhanced by utilizing stereo signals from each of the far endpoints, and squeezing them into lateral angular regions, again, corresponding to the near endpoint video display. The advantage is that now at least some spatial differentiation is available to distinguish talkers located at the same far endpoint.

Three-dimensional (3-D) audio can further enhance the teleconferencing experience. Here, in addition to left/right positioning, up/down and forward/backward is available. There are several ways that free-space 3-D audio can be delivered. The most well known is 5.1-channel surround sound. Another method is 3-D synthesis from stereo using head-related transfer functions (HRTF's) and cross-talk cancellation [26]. The surround sound system is quite robust, but the echo canceler complexity increases as the square of the number of channels. So, a 5-channel system requires 25 times the processing of a monophonic echo canceler. On the other hand, synthesized 3-D audio only requires an echo canceler with the same complexity as a stereo canceler; however, the sweet spot is not as large as with surround sound.

A real-time implementation of a stereo teleconferencing system was developed at Bell Labs in 1998. This will be referred to as the Bell Labs stereo teleconferencing (BLST) system, and its design will be used to exemplify the kind of considerations that are generally regarded.

The elements of a complete stereo teleconferencing echo control system include: adaptive filters for the echo canceler, a scheme for handling double-talk, dynamic suppression, nonlinear processing (NLP), and comfort noise insertion.

The adaptive filters are finite impulse response (FIR) filters that adjust their coefficients in an attempt to predict the echo coming from the room, based on the far-end signal sent to the loudspeakers. However, it is important to inhibit coefficient adaptation under certain circumstances (described below). This is the purpose of the double-talk control function. Dynamic suppression and NLP further attenuate the residual echo that exits the adaptive filters. Because the suppression and NLP change the level of the ambient local room noise they create an annoying modulation of this background noise. To reduce this effect, comfort noise is inserted at the output.

In general, the efficacy of these elements is enhanced if they are implemented within a subband structure. The adaptive filters tend to converge faster, the dynamic suppression and NLP are less obtrusive, and the comfort noise can easily be spectrally shaped to the ambient local room's spectrum. Also, the computational complexity of the adaptive filters decreases linearly with the number of subbands (see Chap. 1).

In addition, some processing of the far-end signal prior to echo control is desirable. In particular, a simple nonlinearity to somewhat "decohere" the left and right channels of the stereo signal helps the adaptive filter to converge (see Chap. 5), and a hard limiter on both channels helps prevent excitation of further nonlinearities in the echo path. Figure 7.1 shows the arrangement of

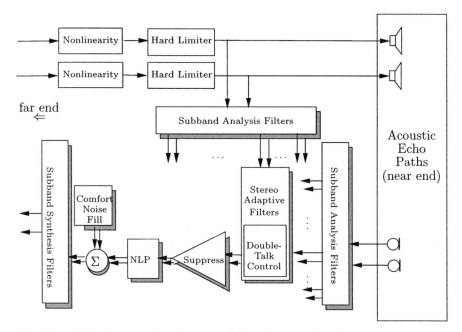

Fig. 7.1. Block diagram of a stereo acoustic echo control system

these elements, which are discussed in detail below. Beginning with the hard limiter, the subject of the following sections proceeds roughly in a clockwise fashion except that the nonlinearity is discussed within the context of the adaptive filter.

7.2 Hard Limiter

The adaptive filter in the echo canceler requires a linear echo path to operate properly since it predicts the echo with a linear filter. Sometimes this requirement is violated when the excitation signal is so large that it becomes clipped in the echo path. For instance, if the microphone is very close to the loudspeaker, a large amplitude excitation signal will cause a large amplitude echo at the microphone which can saturate the microphone pre-amplifier, the amplifier, or the analog-to-digital (A/D) converter. The loudspeaker can also be driven into its nonlinear region of operation by high-amplitude excitation.

The above nonlinearities, in extreme cases, can cause the adaptive filter to diverge. The divergence process is easiest to visualize in the frequency domain. Say the excitation signal is a simple sinusoid. Clipping in the echo path will cause harmonics to appear in the echo signal. The adaptive filter will try to predict these harmonics, but since there is no excitation signal at those frequencies, the filter impulse response may grow without bound in a vain attempt to predict something from nothing.

As far as the echo canceler is concerned, problems due to echo path saturation can be easily avoided by purposely clipping, or hard-limiting, the signal prior to the echo canceler (as in Fig. 7.1). Note that the hard-limiter is placed after the nonlinearity, which may slightly increase the amplitude of the excitation signal. The clipping level should be low enough so that none of the components in the echo path are driven into nonlinearity. At the same time, of course, the clipping level should be kept as high as possible to avoid distorting the excitation signal any more than is absolutely necessary.

7.3 Subband Filter Banks

The subband analysis filterbanks [31] separate the nonlinearly processed far-end signals and near-end signals into M equal-width subbands. The subband synthesis filterbank, on the other hand, combines the M equal-width subband residual error signals into a single full-band error signal.

The subband analysis filterbank can be described by considering only one of the input signals. This is shown in Fig. 7.2. To simplify the notation at this point, let $x(n)$ denote the full-band left excitation signal. The mth subband signal at the subsampled rate is $x_m(k)$. Conceptually, it can be generated by multiplying $x(n)$ with the complex exponential signal, $e^{-j2\pi mn/M}$, to translate the Fourier transform of $x(n)$ down in frequency by m subbands, and then convolving the complex result, $x'_m(n)$, with the (analysis) lowpass filter g_a. Finally the signal is then subsampled by the factor R. The effects of these operations on the spectrum of the subband are shown in Fig. 7.3.

The biggest advantage of subsampling in subbands is that the overall computational rate of the adaptive filters is dramatically lower than at the full-band rate [78], [77] (see Chap. 1). If the number of operations per second to execute the full-band adaptive filter is C_{af} (where the subscript af stands for adaptive filter) MIPS (millions of instructions per second), then the number of operations required to execute the subband adaptive filter is:

$$C_{saf} = \left(C_{cmplx} \frac{M}{2R^2} \right) C_{af} , \qquad (7.1)$$

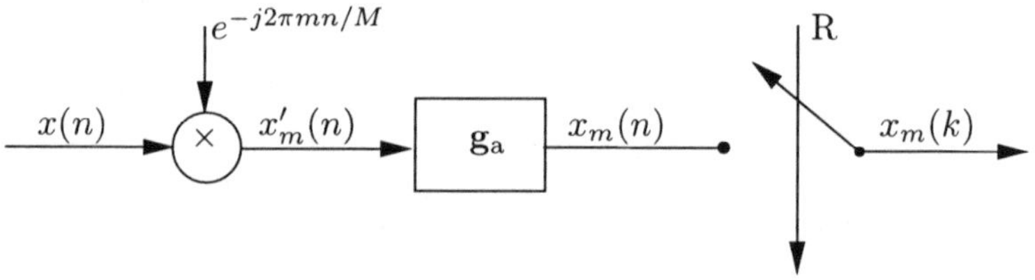

Fig. 7.2. Conceptual block diagram of the mth subband analysis filter

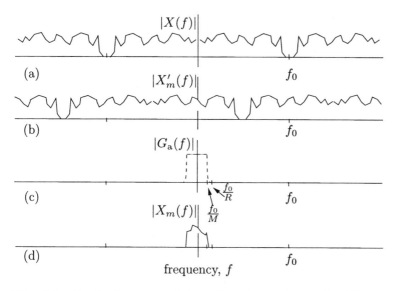

Fig. 7.3. Magnitude spectra of the subband analysis signals of Fig. 7.2

where C_{saf} denotes the subband adaptive filter computation rate and the factor C_{cmplx} takes into account the additional complexity of using complex over real arithmetic (the signals in the subbands are generally complex). There are $M/2$ subbands to process, the adaptive filter lengths are R times shorter in the subbands, and there is R times as much time to do the processing.

In the BLST implementation, the parameters are:

$$C_{\mathrm{cmplx}} = 4 , \tag{7.2}$$
$$M = 128 , \tag{7.3}$$
$$R = 96 , \tag{7.4}$$

making

$$C_{\mathrm{saf}} \approx 0.028 \; C_{\mathrm{af}} . \tag{7.5}$$

So, the subband adaptive filters require a factor of about 36 less computation than the full-band filters. Equation (7.1) takes into account only the complexity of the subband *adaptive* filters and not the subband analysis and synthesis filters. However, the computation associated with the latter is very small in comparison.

7.3.1 Subband Prototype Filters

There are several trade-offs involved in selecting M and R. The relationship between these parameters, the full-band signal, and the subband low-pass

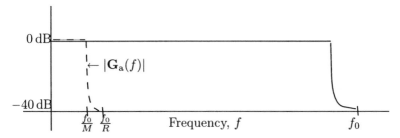

Fig. 7.4. Design of lowpass prototype filter, $\mathbf{G}_a(f)$

prototype filter are shown in Fig. 7.4. The frequency f_0 is the so called *fold-ing frequency* (half the full-band sampling frequency, f_s) of the full-band signal. The solid line denotes the bandwidth of the full-band signal, which is band limited below f_0. The filter, $G_a(f)$ is typically a finite impulse response filter. The magnitude of the frequency response, $|G_a(f)|$, is designed to have large stop-band attenuation at and above f_0/R, the folding frequency of the subbands, thereby preventing aliasing in the subbands.

In applications such as speech coding, some aliasing between subbands is permitted because the aliasing is later eliminated by so-called imaged components during reconstruction [70] (quadrature mirror filters). However, in the case of echo cancellation, the aliased part of the signal causes the adaptive filters within the subbands to converge to a perturbed solution. Also, the aliasing in the signal path doesn't cancel when the subband signals are modified by the adaptive filter. Therefore, aliasing must be strictly controlled.

For good reconstruction, the subband analysis filter's magnitude is -6 dB at f_0/M, the cross-over frequency with the adjacent subband. High quality reconstruction of the near-end signal when it is passed through the analysis and reconstruction filter banks is required so that the near-end talker's voice is transmitted to the far-end without degradation. However, *perfect* reconstruction, a popular topic in the multi-rate literature, is not absolutely necessary in this application. The near-end talker's speech is almost always filtered by the acoustic impulse response of a reverberant room, and room impulse response spectra are anything but flat in magnitude and linear in phase. So, a little distortion in the analysis/synthesis subband filters will most likely pass unnoticed. Nevertheless, filter design techniques that yield near-perfect analysis/synthesis filters should be used. In the BLST system, a slightly modified version of Wackersreuther filters [131] were used. The Wackersreuther design gives flat pass band and stop band magnitude frequency responses with perfect reconstruction between adjacent subbands.

The frequency f_0/R must be greater than f_0/M, or in other words, R must be smaller than M. According to (7.1), though, it is desirable to make R large and M small to keep the adaptive filter's computational complexity low. But as R approaches M, the transition band of the analysis filter must narrow (to avoid aliasing), increasing the length and hence the delay and computational complexity of the analysis filter (with similar effects for the synthesis filter). An analysis filter length of 1000 to 2000 coefficients is not unusual.

The required stop-band attenuation of the subband prototype filter also causes the filter length to be large. This attenuation needs to be rather large to counteract the fact that the other $M - 1$ bands alias into each subband. If the subband signals are considered statistically independent and the same stop band loss of the subband prototype filter is uniformly L_{sb} dB, then the noise in a given subband due to aliasing from the others is (in dB):

$$N_{sb} = L_{sb} + 10 \log_{10}(M - 1) . \tag{7.6}$$

So, if the aliasing noise level is to be kept at say –50 dB, for the example of $M = 128$ and $R = 96$, the required stop-band rejection is $L_{sb} = 71$ dB, which is rather onerous.

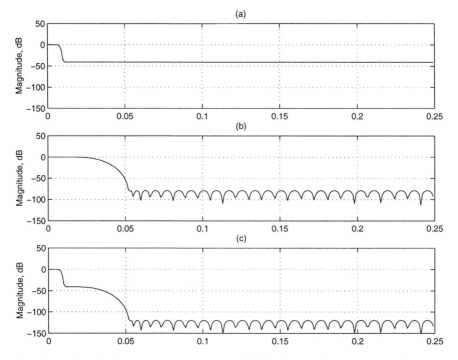

Fig. 7.5. Magnitude frequency responses: (a) Wackersreuther reconstruction prototype lowpass filter, (b) auxiliary lowpass filter, (c) cascade of (a) and (b)

An alternative approach to subband prototype filter design is to drop the idea that the stop-band attenuation be the same for each subband. Rather, two lowpass filters can be cascaded, the first being a perfect reconstruction subband prototype filter and the second being a rather simple and low-cost lowpass filter. This second filter should be linear phase with the flat magnitude frequency response of its pass-band within the transition band of the perfect reconstruction filter. Likewise, the transition band of the second filter should be well into the stop band of the first. This will to a large extent preserve the perfect reconstruction property of the first filter of the cascaded pair. Figure 7.5 shows an example where $M = 128$, $R = 96$, and the stop band attenuation is 120 dB below the passband. Here the Wackersreuther filter has 1490 coefficients, the second filter has 231 coefficients, and the cascade of the two is 1720 coefficients. To achieve similar results with the Wackersreuther design alone requires about 3600 coefficients.

Another advantage of the subband approach is that computational complexity and memory can be saved by simply not processing all of the subbands. This of course compromises the fidelity of the overall system somewhat, but it is a valuable trade-off available to designers. It is typical to not process bands below 150 Hz and those close to the folding frequency. The lower bands are particularly tempting to drop for two reasons. First, the room echo path impulse response is longest in the lower subbands. So, by dropping the them the processing of usually long adaptive filters is avoided. Secondly, ambient room noise is stronger at lower frequencies. In fact, a statistical model often used for simulation and analysis of room noise is called $1/f$ noise (f referring to frequency). As a result, the echo-to-background noise ratio (EBR) is lower in the lower subbands. These facts mean that the lowest bands are usually slower to converge and yield lower echo return loss enhancement (ERLE) than the higher ones.

Of course, the fact that room echo path impulse responses tend to be shorter at higher frequencies can be used to advantage as well. In the BLST system, the subbands above 4 kHz have only half the adaptive filter length of those below. For a full discussion of required echo path lengths in subbands see Diethorn [34].

7.3.2 Subband Analysis Filter Banks

This section discusses efficient implementation of the subband analysis filter banks. Referring back to the block diagram of Fig. 7.2, it is seen that in the mth subband, $x(n)$ is first multiplied by the complex exponential signal $e^{-j2\pi nm/M}$. Denote the mth subband's complex exponential vector as

$$\mathbf{f}_{m,P} = [\, 1 \; e^{-j2\pi m/M} \; e^{-j2\pi 2m/M} \; \cdots \; e^{-j2\pi(P-1)m)/M} \,]^T$$

$$= [\, \mathbf{I}_M \; \cdots \; \mathbf{I}_M \,]^T \mathbf{f}_{m,M} \,, \tag{7.7}$$

where we have assumed that P, the length of the subband analysis lowpass prototype filter, is a multiple of M so that (by the M-perodicity) $f_{m,P}$ can be subdivided into P/M sub-vectors, $f_{m,M}$, and $\mathbf{I}_M$ is the $M \times M$ identity matrix.

Define the input signal vector

$$\mathbf{x}(n) = [\,x(n)\ x(n-1)\ \cdots\ x(n-P+1)\,]^T\,.$$

Then the mth subband signal prior to subsampling can be written as

$$
\begin{aligned}
x_m(n) &= \mathbf{g}_a^T \mathrm{diag}\{\mathbf{f}_{m,P}\}\mathbf{x}(n) \\
&= \mathbf{f}_{m,P}^T \mathrm{diag}\{\mathbf{g}_a\}\mathbf{x}(n) \\
&= \mathbf{f}_{m,M}^T[\mathbf{I}_M\ \cdots\ \mathbf{I}_M]\mathrm{diag}\{\mathbf{g}_a\}\mathbf{x}(n) \\
&= \mathbf{f}_{m,M}^T \mathbf{G}_a\mathbf{x}(n)\,,
\end{aligned}
\tag{7.8}
$$

where $\mathrm{diag}\{\mathbf{f}_{m,P}\}$ stands for the diagonal matrix made up of the elements of the vector $\mathbf{f}_{m,P}$ and the $M \times P$ sparse matrix, $\mathbf{G}_a$, is defined as

$$
\mathbf{G}_a =
\begin{bmatrix}
g_{a,0} & 0 & g_{a,M} & 0 & g_{a,P-M} & 0 \\
& \ddots & & \ddots & \cdots & \ddots \\
0 & g_{a,M-1} & 0 & g_{a,2M-1} & 0 & g_{a,P-1}
\end{bmatrix}.
\tag{7.9}
$$

Recognizing that the $M \times M$ discrete Fourier transform (DFT) matrix can be written as

$$
\mathbf{F}_M =
\begin{bmatrix}
\mathbf{f}_{0,M}^T \\
\vdots \\
\mathbf{f}_{m,M}^T \\
\vdots \\
\mathbf{f}_{M-1,M}^T
\end{bmatrix},
\tag{7.10}
$$

it is easy to see that the vector of subband signals from subband 0 to $M-1$ can be expressed as

$$
\boldsymbol{\chi}(n) =
\begin{bmatrix}
x_0(n) \\
\vdots \\
x_m(n) \\
\vdots \\
x_{M-1}(n)
\end{bmatrix}
=
\begin{bmatrix}
\mathbf{f}_{0,M}^T \\
\vdots \\
\mathbf{f}_{m,M}^T \\
\vdots \\
\mathbf{f}_{M-1,M}^T
\end{bmatrix}
\mathbf{G}_a\mathbf{x}(n) = \mathbf{F}_M\mathbf{G}_a\mathbf{x}(n)\,.
\tag{7.11}
$$

Since the elements of $\boldsymbol{\chi}(n)$ are band limited, they may be subsampled, meaning that (7.11) need only be calculated once every R sample periods to obtain the subsampled subband signals, $\boldsymbol{\chi}(k)$.

Figure 7.6 is a block diagram representing the processing of (7.11). The far-end samples $x(n)$ are shifted into a tapped delay line at the full-band rate indicated by the time index, n. Each delay is then subsampled by the factor R. Another way to think of it is that R samples of data are shifted into

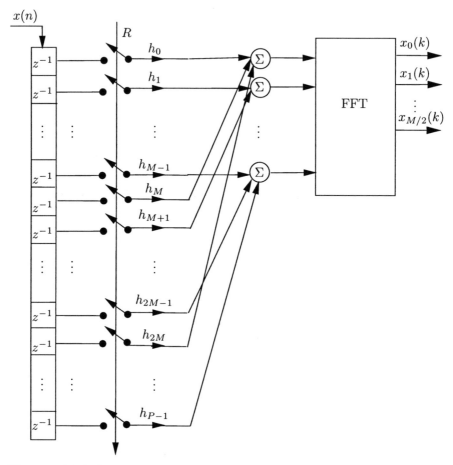

Fig. 7.6. Block diagram of the subband analysis filter bank

the delay line between each subsample period indexed by k. The delay line outputs are multiplied by the prototype filter coefficients. The sparse nature of $\mathbf{G_a}$ is exploited in the next step where the M inputs to the fast Fourier transform (FFT) are computed. For example, since the first row of $\mathbf{G_a}$ consists of h_0, followed by $M-1$ zeros, followed by h_M, followed by $M-1$ zeros, etc., the first element of the FFT input is computed by only summing the non-zero terms. The remaining FFT inputs are computed similarly following the structure of $\mathbf{G_a}$. When the FFT is computed, only the first $1 + M/2$ terms are kept since the remainder are redundant; since the input signal is real, the FFT outputs above and below $M/2$ are conjugate symmetric about $M/2$. The subband signals of the FFT outputs are generally complex.

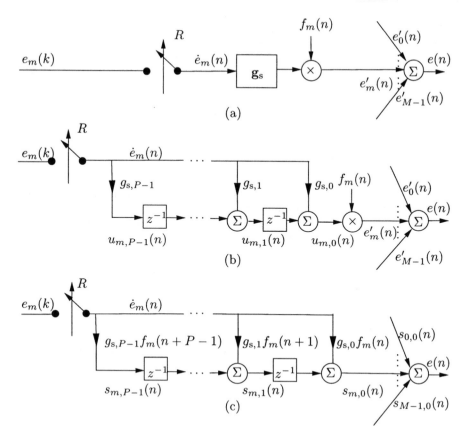

Fig. 7.7. Subband synthesis filter. (a) Generic structure. (b) Structure using a type II FIR filter. (c) Equivalent to (b)

7.3.3 Subband Synthesis Filter Bank

A generic structure of the synthesis filterbank, focusing on the mth subband, is shown in Fig. 7.7. The signals being "reconstructed" are the generic subband error signals, $e_m(k)$, which are comprised as the outputs of the subband NLPs and comfort noise generators as shown in Fig. 7.1. In Fig. 7.7(a), $e_m(k)$ is upsampled by inserting $R-1$ zeros between each sample of $e_m(k)$, creating $\dot{e}_m(n)$. This causes images of $e_m(k)$ to repeat in the frequency domain. Next, $\dot{e}_m(n)$ is lowpass filtered with the reconstruction filter, $\mathbf{g}_s$, to eliminate the imaged components. Usually one choses the filter $\mathbf{g}_s$ as a time-reversed version of $\mathbf{g}_a$ used in the analysis filter banks in order to obtain an overall linear phase response. Thus:

$$\begin{aligned}
\mathbf{g}_s &= \begin{bmatrix} g_{s,0} \ g_{s,1} \ \cdots \ g_{s,P-2} \ g_{s,P-1} \end{bmatrix}^T \\
&= \begin{bmatrix} g_{a,P-1} \ g_{a,P-2} \ \cdots \ g_{a,1} \ g_{a,0} \end{bmatrix}^T .
\end{aligned} \tag{7.12}$$

The result is then shifted up in frequency by m subbands by multiplication with the complex exponential signal,

$$f_m(n) = e^{j2\pi mn/M} , \tag{7.13}$$

to obtain $e'_m(n)$. Finally, this is summed with the other similarly processed subband signals to obtain the reconstructed signal

$$e(n) = \sum_{m=0}^{M-1} e'_m(n) . \tag{7.14}$$

Figure 7.7(b) shows a subband synthesis filter where a direct form II finite impulse response (FIR) filter is used. The state variables of the mth subband filter, $u_{m,p}(n)$, $p = 0, ..., P - 1$, are the input to the left-most delay and the outputs of each of the $P - 1$ summers in the filter. They are represented by the vector

$$\mathbf{u}_m(n) = \begin{bmatrix} u_{m,0}(n) \\ u_{m,1}(n) \\ \vdots \\ u_{m,P-2}(n) \\ u_{m,P-1}(n) \end{bmatrix} = \begin{bmatrix} g_{s,0} \\ g_{s,1} \\ \vdots \\ g_{s,P-2} \\ g_{s,P-1} \end{bmatrix} \dot{e}_m(n) + \begin{bmatrix} u_{m,1}(n-1) \\ u_{m,2}(n-1) \\ \vdots \\ u_{m,P-1}(n-1) \\ 0 \end{bmatrix} . \tag{7.15}$$

Define an alternative state vector,

$$\mathbf{s}_m(n) = \begin{bmatrix} s_{m,0}(n) \\ s_{m,1}(n) \\ \vdots \\ s_{m,P-1}(n) \end{bmatrix} = \begin{bmatrix} f_m(n)u_{m,0}(n) \\ f_m(n+1)u_{m,1}(n) \\ \vdots \\ f_m(n+P-1)u_{m,P-1}(n) \end{bmatrix} . \tag{7.16}$$

Multiply (7.15) from the left by $\text{diag}\{ f_m(n) \; f_m(n+1) \cdots f_m(n+P-1) \}$ and use equation (7.16) to obtain

$$\begin{bmatrix} s_{m,0}(n) \\ s_{m,1}(n) \\ \vdots \\ s_{m,P-2}(n) \\ s_{m,P-1}(n) \end{bmatrix} = \begin{bmatrix} f_m(n)g_{s,0} \\ f_m(n+1)g_{s,1} \\ \vdots \\ f_m(n+P-2)g_{s,P-2} \\ f_m(n+P-1)g_{s,P-1} \end{bmatrix} \dot{e}_m(n) + \begin{bmatrix} s_{m,1}(n-1) \\ s_{m,2}(n-1) \\ \vdots \\ s_{m,P-1}(n-1) \\ 0 \end{bmatrix} . \tag{7.17}$$

Now, expressly define the lower $P - 1$ elements of $\mathbf{s}_m(n)$ as

$$\tilde{\mathbf{s}}_m(n) = \begin{bmatrix} s_{m,1}(n) \\ \vdots \\ s_{m,P-1}(n) \end{bmatrix} . \tag{7.18}$$

Then (7.17) can be rewritten as

$$
\mathbf{s}_m(n) = \begin{bmatrix} g_{s,0} & & & 0 \\ & g_{s,1} & & \\ & & \ddots & \\ 0 & & & g_{s,P-1} \end{bmatrix} \mathbf{f}_{m,P}^* \dot{e}_m(n) + \begin{bmatrix} \tilde{\mathbf{s}}_m(n-1) \\ 0 \end{bmatrix} , \tag{7.19}
$$

where $\mathbf{f}_{m,P}$ is defined in (7.7) and $*$ denotes the complex conjugate. This is the structure shown in Fig. 7.7(c).

Equation (7.19) can alternatively be expressed as

$$
\mathbf{s}_m(n) = \mathbf{G}_s \mathbf{f}_{m,M}^* \dot{e}_m(n) + \begin{bmatrix} \tilde{\mathbf{s}}_m(n-1) \\ 0 \end{bmatrix} , \tag{7.20}
$$

where $\mathbf{G}_s$ is the sparse matrix,

$$
\mathbf{G}_s = \begin{bmatrix} g_{s,0} & & & 0 \\ & \ddots & & \\ 0 & & g_{s,M-1} & \\ g_{s,M} & & 0 & \\ & \ddots & & \\ 0 & & g_{s,2M-1} & \\ & \vdots & & \\ g_{s,P-M} & & 0 & \\ & \ddots & & \\ 0 & & g_{s,P-1} \end{bmatrix} . \tag{7.21}
$$

Now, sum the state variables across subbands to get

$$
\mathbf{s}(n) = \sum_{m=0}^{M-1} \mathbf{s}_m(n) = \mathbf{G}_s \sum_{m=0}^{M-1} \mathbf{f}_{m,M}^* \dot{e}_m(n) + \sum_{m=0}^{M-1} \begin{bmatrix} \tilde{\mathbf{s}}_m(n-1) \\ 0 \end{bmatrix} . \tag{7.22}
$$

Define the vector of upsampled subband signals as

$$
\mathcal{E}(n) = \begin{bmatrix} \dot{e}_0(n) \\ \dot{e}_1(n) \\ \vdots \\ \dot{e}_{M-1}(n) \end{bmatrix} . \tag{7.23}
$$

Recognizing that

$$
\mathbf{F}_M^{-1} \mathcal{E}(n) = \sum_{m=0}^{M-1} \mathbf{f}_{m,M}^* \dot{e}_m(n) \tag{7.24}
$$

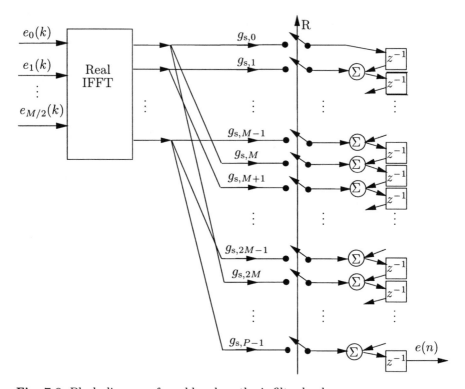

Fig. 7.8. Block diagram of a subband synthesis filter bank

[see (7.10)] and defining

$$\tilde{s}(n-1) = \sum_{m=0}^{M-1} \tilde{s}_m(n-1) ,\tag{7.25}$$

(7.22) can be written as

$$\mathbf{s}(n) = \mathbf{G}_s \mathbf{F}_M^{-1} \mathcal{E}(n) + \begin{bmatrix} \tilde{s}(n-1) \\ 0 \end{bmatrix} .\tag{7.26}$$

Since $\mathcal{E}(n)$ is only non-zero once every R samples, the only change in $\mathbf{s}(n)$ for $R-1$ of those sample periods is that the components shift vertically. Obviously, then, the computational work of the reconstruction need only be done once every R samples, i.e., when $\mathcal{E}(n)$ is non-zero. After each computation, the R new samples of the reconstructed signal are shifted out of $\mathbf{s}(n)$.

An efficient implementation of (7.26) is shown in Fig. 7.8. First the inverse FFT (IFFT) is performed on the subband signals. Figure 7.8 shows that only bands 0 through $M/2$ are used as inputs, which differs from (7.26) where all M subbands are used. The reason for this is that the upper $M/2 - 1$ bands

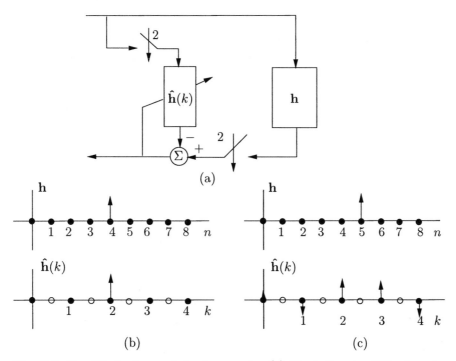

Fig. 7.9. Simplified sub-sampled echo canceler. (a) Block diagram. Filter impulse responses (b) when an ideal echo path impulse falls on a subsample, (c) when the echo path impulse does not fall on a subsample

are merely complex conjugate representations of the lower bands, 1 through $M/2$. The real IFFT implementation easily takes this into account.

At the output of the IFFT, the multiplication with the sparse matrix $\mathbf{G_s}$ is implemented. Its sparse structure is exploited such that each coefficient of the reconstruction filter multiplies only one of the IFFT outputs prior to being upsampled and summed into the state vector. The effect of the upsampling is that the column of summers really only do a true sum once every R samples. During the remaining $R - 1$ periods, the summer/memory bank merely operates as a shift register.

7.4 Non-Causal Taps

The subsampling operation in the subband filterbanks gives rise to the interesting phenomena of non-causal coefficients in the adaptive filter [77], [78] (see Chap 1). For the adaptive filters in the subbands to fully represent the full band echo path impulse response, some coefficients prior to the advent of the echo path in the full band are required in the subbands. At first, this may seem counter intuitive, but a simple example may clarify the phenomenon.

Assume that there is no near-end signal and let the excitation and echo signals be band limited to the extent that the subband filter effects can be neglected. To further simplify matters, assume that the subsampling factor $R = 2$. Figure 7.9(a) shows the system. (For simplicity, this section addresses this issue in terms of a monophonic canceler, the extension to stereo case being obvious.) Without loss of generality, let the subsampling occur in such a manner that only the even samples (e.g., $n = 0, 2, 4, ...$) are saved and the odd ones are discarded. If the echo path impulse response was a simple delay,

$$h = \delta(n - d) , \tag{7.27}$$

where $\delta(n)$ is the discrete impulse function and the delay of the echo path, d, is an even number of full-band samples, then the converged adaptive filter, $\hat{h}(n)$, will have a single non-zero tap at delay, $k = d/2$ [see Fig. 7.9(b)]. But, what happens if the the delay in the echo path is an *odd* number of full band taps? The converged adaptive filter must then interpolate (with a sinc function) at the subsampled rate before and after the time instant $k = d/2 + 1/2$ [see Fig. 7.9(b)]. The filter coefficients prior to the onset of the echo path impulse are called the *non-causal* taps. Typically, allowing for 5 to 10 non-causal coefficients in the *subbands* is sufficient for most acoustic echo cancellation applications. This amounts to adding delay somewhere between the excitation signal input, $x(n)$, and where the estimated echo is subtracted from the echo. Sometimes the direct path acoustic delay between the loudspeaker and the microphone is sufficient to supply the necessary additional coefficients. However, when this delay is unknown or variable, some delay must be added between the microphone and the echo canceler to insure good performance under all conditions.

7.5 Stereo Acoustic Echo Canceler

The stereo acoustic echo canceler consists of a two-channel adaptive filter and some form of double-talk control. The general form of a stereo adaptive filter is shown in Fig. 7.10. The signals from the far-end are first processed by a simple nonlinearity to decrease the coherence between the left and right channels (see Chap. 5). (For simplicity, we do not show here the hard limiter in Fig. 7.1, the purpose of which is discussed in Sect. 7.2.) Then, the left signal, $x_1(n)$, is fed into the left speaker as well as adaptive filters, $\hat{h}_{11}$ and $\hat{h}_{21}$. The right far-end signal, $x_2(n)$, is fed into the right speaker, $\hat{h}_{12}$ and $\hat{h}_{22}$. The outputs of $\hat{h}_{11}$ and $\hat{h}_{12}$, summed together, form the prediction of the left microphone signal, $\hat{y}_1(n)$. This prediction is subtracted from the actual microphone signal, $y_1(n)$, to obtain the left error signal, $e_1(n)$, which is the left-channel return signal, and is also used by the adaptation algorithm to adjust the filter coefficients. The right microphone's canceler operates similarly.

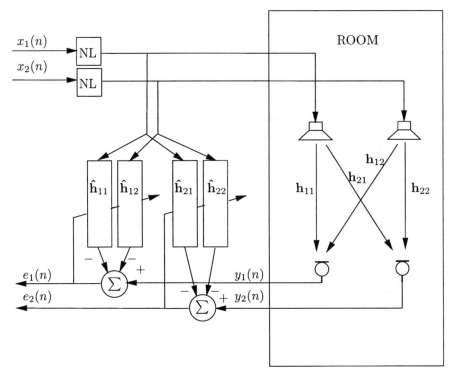

Fig. 7.10. Block diagram of a stereo adaptive filter

Even with the nonlinearities on the excitation signals, the condition number on the stereo correlation matrix (due to the common acoustic source) is normally very high. This causes NLMS-based adaptive filters to converge and track too slowly (even within the subband framework) to effectively eliminate echo. For this reason, a multi-channel recursive least-squares (RLS) algorithm is used. RLS algorithms have fast convergence properties especially in the case where the excitation covariance matrix has a high condition number (see Chaps. 3 and 5). Computationally efficient versions of the RLS algorithm are also available, which are "fast" in a computational sense. Fast RLS (FRLS) [10] algorithms exploit shift invariance and the rank-one updating properties of the covariance matrix (when the performance index is an exponentially weighted sum of squared errors). The stereo FRLS adaptive filter is more complex to implement than the stereo NLMS [8]: FRLS requires roughly $32L$ multiply/adds per sample period, as compared to roughly $8L$ multiply/adds for NLMS. As for memory, FRLS requires $12L$ locations while NLMS requires $6L$ memory locations.

The main drawback to FRLS algorithms is that they occasionally experience exponential divergence [116]. However, as explained below, double-talk control can greatly alleviate this problem. Divergence in FRLS adaptive fil-

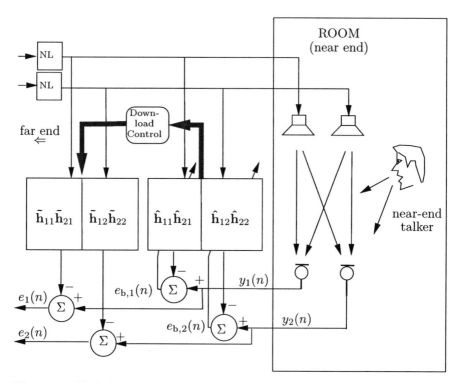

Fig. 7.11. Block diagram of two-path double-talk control technique

ters is often predicted (or at least quickly detected) by so-called "rescue variables" [29]. When the rescue variables get smaller than a certain threshold, the adaptive filter is restarted.

One excellent method of double-talk control is the so-called two-path echo canceler [104] where there are two sets of filters, background and foreground, that predict the echo (see Fig. 7.11). The background filters, $\hat{\mathbf{h}}_{pq}$, $p, q = 1, 2$, always adapt their coefficients, even during double-talk. The foreground filters, $\tilde{\mathbf{h}}_{pq}$, periodically receive their coefficients from the background filters when a series of tests indicate that it is favorable to do so. Only the error signals of the foreground filters are returned to the far end, so allowing the background filters to diverge during double-talk does not affect the perceived performance of the system. In fact, the background filters are sometimes called *sacrificial* filters.

The added complexity of the two-path method is quite small. In the stereo FLRS case, it requires only $4L$ additional multiply-adds compared to the $32L$ already used for the adaptive filters, requiring only a 12.5 percent increase. The memory increase is not quite so insignificant, requiring an additional 33 percent.

The two-path method not only protects the user from hearing the effects of bad coefficients due to double-talk, but also from the occasional FRLS algorithm divergence. When the FRLS coefficients diverge, they do so very quickly, usually in just a few samples. This is easy for the download tests to detect, so users almost never notice any effects due to FRLS divergence.

The key to good two-path method performance lies in the definitions of the download tests. In the BLST implementation, the tests are performed in each subband once every 12 ms. The tests in each subband are phased so that half the subbands are tested and downloaded in one frame, and the other subbands in the next frame. This distributes the testing and downloading computation (which involves memory transfers) relatively evenly in time.

In [104], three download tests are specified. First define the following frame-based energy measures:

- σ_y^2 is the sum of the squared $y_1(k)$ and $y_2(k)$'s over the frame length.
- σ_x^2 is the sum of the squared $x_1(k)$ and $x_2(k)$'s over the frame length.
- σ_e^2 is the sum of the squared $e_1(k)$ and $e_2(n)$'s over the frame length.
- $\sigma_{e_b}^2$ is the sum of the squared $e_{b,1}(k)$ and $e_{b,2}(k)$'s over the frame length.

The download tests are then:

1. Is $\sigma_y^2 < T_1\sigma_x^2$? That is, is the microphone signal energy smaller than expected given the excitation signal and the known echo path gain, T_1?
2. Is $\sigma_{e,b}^2 < T_2\sigma_y^2$? That is, is the background error energy smaller than the microphone energy by a factor of T_2 (typically set to about 12 dB)?
3. Is $\sigma_{e,b}^2 < T_3\sigma_e^2$? That is, is the background error energy smaller than the foreground error energy by a factor of T_3 (typically set to about 3 dB)?

When all three tests are passed in three consecutive frames, then the background coefficients are downloaded. The thinking behind these tests is as follows.

First, in a double-talk situation, speech comes from both the far-end and near-end talkers. If there is a good estimate of the echo path gain, T_1, the microphone level can be well approximated. If the microphone energy is above that level, then someone on the near-end must be talking. The problem with this test is that it relies on a confident measure of T_1, which can change with the acoustic environment or with gain settings in the electronics. So, one must be careful not to place too much confidence in this test. Therefore, preferring to error on the side of fast tracking, it is often a good idea to make T_1 somewhat larger than the gain that is actually measured.

The second test is justified by the following argument. When the near-end talker is active, the background filter coefficients will diverge from the true echo path impulse response as it tries to drive its error signal to zero by attempting to decorrelate (implicitly, at least) the excitation and microphone signals. Since the far-end and near-end signals are uncorrelated, the background adaptive filter will fail to make the error signals much smaller than the microphone signals. The foreground filter will also not predict the near-end signal (which is perfectly fine since the purpose of the canceler is certainly not to cancel the near-end speech), so the ratio of error to microphone signal energy will also be near 0 dB.

The third test assures that background coefficients that give greater error energy than those in the foreground are not downloaded. The threshold parameter of T_3 being a little smaller than 0 dB provides a small amount of hysteresis.

In [104], all the signals were pre-filtered with a low-order prediction filter derived from the excitation signal. Since the adaptive filters in the BLST implementation already operate in narrow subbands and hence are fairly white anyway, the prediction filtering was dispensed with without loss of performance.

A few more tests improve the performance of the two-path method considerably. These are:

4. Is $\sigma_y^2 > T_{\text{SNR}}\sigma_{\text{AN}}^2$? That is, is the microphone signal energy larger than ambient noise level in the room by a factor of T_{SNR}?
5. Is $\xi_m(k) > T$? That is, is the normalized excitation/echo correlation measure, $\xi_m(k)$, (see Chap. 6) greater than the threshold, T?

In Test 4, downloading is only considered when the microphone signal energy is a certain level above the ambient room noise energy measurement, σ_{AN}^2. This avoids the situation of making download decisions when there is very little energy in (and hence little knowledge of) the echo path.

In Test 5, the normalized correlation measure [16] discussed in Chap. 6 is tested. This measure varies from 0 to 1 when the excitation and microphone signals are uncorrelated or correlated, respectively. Of course if only echo is present at the microphone, $\xi_m(k)$ will be very close to 1. If, however, the near-end talker is active, the microphone signal will be uncorrelated with the excitation (unless the two users are singing in unison) and $\xi_m(k)$ will be small. Normally this is a very expensive measure to implement, but since FRLS is already being used for the adaptive filters, most of the computations are already done (see Chap. 6).

7.6 Dynamic Suppression

In a perfect world, suppression and subsequent nonlinear processing (NLP) (discussed in the next section) would not be necessary in an echo control system. One would only need an echo canceler to achieve echo free full-duplex conversations. However, even with the fastest tracking adaptive filters working with perfect double-talk control techniques, the residual echo can be quite strong. For instance, if the echo path changes dramatically during a long double-talk period when adaptation of the foreground filters is inhibited, the

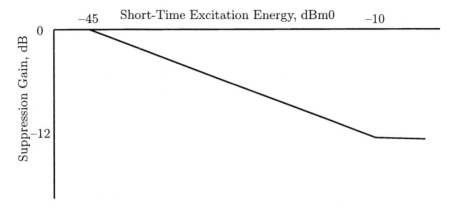

Fig. 7.12. Suppression gain as a function of short-time excitation energy

echo return loss enhancement (ERLE) can get close to 0 dB. So, residual echo control is absolutely required in any practical audio teleconferencing system. The difficult task is to balance the conflicting requirements of full-duplex operation and echo control.

Low levels of excitation (far-end speech) require relatively little attenuation since they produce low residual echo levels, while high excitation levels produce correspondingly high residual echo levels and require greater attenuation so as not to be perceived at the far end. In keeping with this philosophy, the suppressor attenuates the echo canceler's residual echo signal according to a short-time energy estimate of the excitation signal as shown in Fig. 7.12. When the excitation energy estimate is below –45 dBm0 (in telecommunications, +3 dBm0 corresponds to the maximum level of the signal) the suppressor gain is 0 dB. Above –45 dBm0, the suppression gain decreases linearly in the log domain until it reaches –12 dB, corresponding to a short-time excitation energy of –10 dBm0. Above –10 dBm0, the attenuation level remains at –12 dB.

7.7 Nonlinear Processing

Following the adaptive filter and the dynamic suppressor is the nonlinear processor (NLP). Often, this consists of a center clipper, which sets its gain to zero when the level of the input falls below a time-varying threshold, $T_{m,C}(k)$, but otherwise has unity gain. This is shown in Fig. 7.13(a). Typically, in echo cancelers, the threshold is set to a level of about 20 dB below a short-time magnitude estimate of the excitation signal,

$$\sigma_{x,m}(k) = \lambda_{ST}\sigma_{x,m}(k) + (1 - \lambda_{ST})\left[|x_{m,1}(k))| + |x_{m,2}(k)|\right] , \qquad (7.28)$$

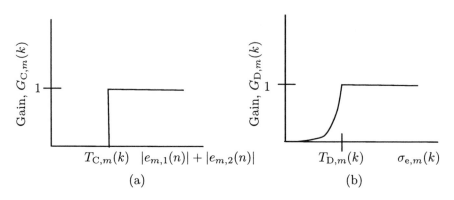

Fig. 7.13. Gain plot of (a) standard center clipper (b) modified Diethorn NLP

where λ_{ST} corresponds to a forgetting factor commensurate with the general reverberation time of the room. A short-time error magnitude signal, $\sigma_{e,m}(k)$ is calculated in the same way. The center clipping threshold is then

$$T_{C,m}(k) = \gamma_0 \sigma_{x,m}(k) , \tag{7.29}$$

where γ_0 is the nominal suppression threshold (e.g. $-20\,$dB).

The general idea is that the echo canceler will decrease the echo level by 12 to 20 dB, and the echo suppressor will reduce it an additional 65 dB or so, depending on the excitation level. By the time the suppressed echo reaches the center clipper it should be down by at least 18 to 24 dB from the excitation level. Then the center clipper will remove it entirely. Near-end speech however, being much larger than the suppressed residual echo, will pass relatively unscathed. Obviously, it is desirable to keep the center-clipping threshold as small as possible to avoid corrupting near-end speech too much.

The BLST system, however, uses a more advanced form of NLP which is a slightly modified version of one suggested by Diethorn [35]. While it is similar to the center clipper described above, there are a few important differences. First, the time varying threshold includes an independently measured echo path gain for each subband, G_m,

$$T_{D,m}(k) = G_m \gamma_0 \sigma_{x,m}(k) . \tag{7.30}$$

Then, the NLP gain rule [shown in Fig. 7.13(b)] is

$$G_{D,m}(k) = \begin{cases} \left[\dfrac{\sigma_{e,m}(k)}{T_{D,m}(k)}\right]^2 , & \sigma_{e,m}(k) < T_{D,m}(k) \\ 1 , & \text{otherwise} \end{cases} , \tag{7.31}$$

where $\sigma_{e,m}(k)$ is the estimated standard deviation of total error in the mth subband, defined similar to (7.28).

If G_m is greater than $0\,\mathrm{dB}$, it may boost the residual echo signal level to a point where it can be perceived by the far-end user. To prevent this, $T_{\mathrm{D},m}(k)$ increases, causing attenuation to be applied at a level higher than in the center clipper above. On the other hand, if G_m is low, the residual will likewise be low and attenuation can be applied at a lower level, disturbing the signal less.

Note, that in the center clipper case, the gain is a function of the error magnitude whereas in the Diethorn NLP it is a function of the smoothed error magnitude. Also, the gain does not suddenly drop to zero as in the center clipper case, but descends more smoothly. Both of these changes cause the NLP gain to change less abruptly than in a standard center clipper.

7.8 Comfort Noise Fill

The processes of suppression and NLP will vary the level of the background noise during the conversation. This so-called "noise pumping" can be quite annoying to the listener on the far end. Adding "comfort noise" to the NLPed and suppressed signal can alleviate this problem. The added noise should be close in amplitude to the background noise from the near end, implying that this background noise must be measured. Fortunately, since the system operates in subbands, a colored background noise spectrum can be accommodated: estimating the noise in each subband and imitating it with a white noise generator at that level will reproduce a similar sounding signal when the subbands are reconstructed in the synthesis filter bank. To be specific, the near-end background noise level in a subband is the level of the signal in that subband when there is no echo or near-end talking present. Since it is easy to detect the presence of the echo by monitoring the short-time energy in the excitation, the main difficulty is estimating the background noise in the presence of the near-end talker since there is no independent knowledge of when the near-end talker is active. One way around this problem is to use so-called *minimum statistics* [90]. The idea is to periodically sample short-time background noise energy estimates and put them in a tapped delay line, say, of length P. The background noise level is just the minimum estimate over the past, say, 30 seconds.

The only difficulty with the minimum statistics method, as presented so far, is that the P-length delay line of noise level estimates must be searched every sample period. Performing a full search only once every frame of K sample periods can mitigate this complexity. In the periods between full searches, the incoming energy estimates are compared against the latest declared minimum and if the new estimate is found to be lower, it is declared to be the new minimum. This reduces the search complexity by a factor of K at the insignificant price of causing a small amount of jitter over the analysis window. Further complexity reduction can be achieved by having the tapped delay line hold the minimum of the K newest estimates at the end of each frame. Now to perform the analysis over P sample periods, only P/K memory locations are required, and an additional savings of a factor of K is realized.

7.9 Conclusions

This chapter has described many of the more practical aspects of a stereo teleconferencing system implementation. It is based on a system built in the Acoustics and Speech Research Department of Lucent Technologies, Bell Laboratories in Murray Hill, New Jersey. Each of the main subsystems of the system was discussed: hard limiting, the stereo nonlinearity, the subband analysis and synthesis filters, non-causal delay, adaptive filters, double-talk control, dynamic suppression, NLP, and comfort noise fill.

Experience with this system has shown that stereo teleconferencing allows users to interact with each other across a connection more like they would if they were together in the same room. It accomplishes this by allowing the users to fully utilize the human binaural hearing system they were born with.

8. General Derivation of Frequency-Domain Adaptive Filtering

8.1 Introduction

Adaptive filters [60] play an important role in echo cancellation because we need to identify and track unknown and time-varying channels [24]. There are roughly two classes of adaptive algorithms. One class includes filters that are updated in the time domain, sample-by-sample in general, like the classical least mean square (LMS) [134] and recursive least-squares (RLS) [4], [66] algorithms. The other class contains filters that are updated in the frequency domain, block-by-block in general, using the fast Fourier transform (FFT) as an intermediary step. As a result of this block processing, the arithmetic complexity of the algorithms in the latter category is significantly reduced compared to time-domain adaptive algorithms. Use of the FFT is appropriate to the Toeplitz structure, which results from the time-shift properties of the filter input signal. Consequently, deriving a frequency-domain (FD) adaptive algorithm is just a matter of rewriting the time-domain error criterion in a way that Toeplitz and circulant matrices are explicitly shown.

Since its first introduction by Dentino *et al.* [33], adaptive filtering in the frequency domain has progressed rapidly, and different sophisticated algorithms have since been proposed. Ferrara [45] was the first to elaborate an efficient frequency-domain adaptive filter algorithm (FLMS) that converges to the optimal (Wiener) solution. Mansour and Gray [88] derived an even more efficient algorithm, the *unconstrained* FLMS (UFLMS), using only three FFT operations per block instead of five for the FLMS, with comparable performance [84]. However, a major handicap with these structures is the delay introduced between input and output. Indeed, this delay is equal to the length of the adaptive filter L, which can be considerable for some applications like acoustic echo cancellation (AEC) where the number of taps can be easily a thousand or more. A new structure called *multi-delay filter* (MDF), using the classical overlap save (OLS) method, was proposed in [123], [6] and generalized in [7], [103], [13] where the block length N was made independent of the filter length L; N can be chosen as small as desired, so as to limit the block delay. Although from a complexity point of view, the optimal choice is $N = L$, using smaller block sizes ($N < L$) in order to reduce the delay is still more efficient than time-domain algorithms. A more general scheme based on weighted overlap and add (WOLA) methods, the *generalized multi-delay filter* (GMDFo), was proposed in [102], [108], where o is the overlap factor. The settings $o > 1$ appear to be very useful in the context of adaptive filtering, since the filter coefficients can be adapted more frequently (every N/o samples instead of every N samples in the standard OLS scheme). Thus, this structure introduces one more degree of freedom, but the complexity is increased by roughly a factor o. Taking the block size in the MDF as large as the delay permits, will increase the convergence rate of the algorithm, while taking the overlap factor greater than 1 will increase the tracking ability of the algorithm. Note that a delayless multirate structure has also been proposed which may be more efficient for N small [93].

As we can see, there are many frequency-domain adaptive algorithms in the literature, but none of them has been derived rigorously. Reference [114] gives an overview of frequency-domain algorithms for the case $N = L$. In this chapter, we propose a new theory on how to develop a whole class of adaptive filters in the frequency domain from a recursive least-squares criterion, with a block size independent of the length of the adaptive filter.

The organization of this chapter is as follows. In Sect. 8.2, we propose a frequency-domain recursive least-squares criterion from which the so-called normal equation is derived. Then, from this normal equation, we deduce an exact adaptive algorithm in the frequency domain that we can write in different ways. In the last part of this section, we study the convergence of this algorithm. In Sect. 8.3, we give a very useful approximation, deduce several well-known algorithms, and give hints on the choice of some parameters (to make the algorithms behave nicely in practice) such as the exponential window, regularization, and adaptation step size. Section 8.4 shows a rigorous link between two classical algorithms: the MDF and the affine projection algorithm (APA). In Sect. 8.5, we briefly generalize some of these ideas to the multichannel case. Finally, we give our conclusions in Sect. 8.6.

8.2 General Derivation of FD Adaptive Algorithms

The first part of this section shows how to write a block recursive least-squares criterion in the frequency domain, with a block size independent of the length of the adaptive filter. When the criterion is rigorously defined, the adaptive algorithm follows immediately.

8.2.1 Criterion

In the context of system identification (see Fig. 8.1), the error signal at time n between the system and model filter outputs is given by

$$e(n) = y(n) - \hat{y}(n) , \tag{8.1}$$

where

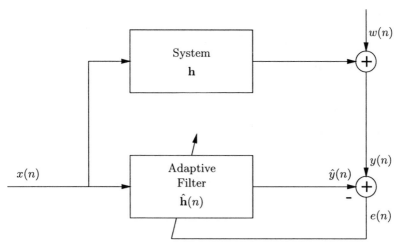

Fig. 8.1. System identification

$$\hat{y}(n) = \hat{\mathbf{h}}^T \mathbf{x}(n) \tag{8.2}$$

is an estimate of the output signal $y(n)$,

$$\hat{\mathbf{h}} = \begin{bmatrix} \hat{h}_0 \ \hat{h}_1 \ \cdots \ \hat{h}_{L-1} \end{bmatrix}^T$$

is the model filter,

$$\mathbf{x}(n) = \begin{bmatrix} x(n) \ x(n-1) \ \cdots \ x(n-L+1) \end{bmatrix}^T$$

is a vector containing the last L samples of the input signal x, and superscript T denotes transpose of a vector or a matrix. We assume that:

$$y(n) = \mathbf{h}^T \mathbf{x}(n) + w(n) , \tag{8.3}$$

where

$$\mathbf{h} = \begin{bmatrix} h_0 \ h_1 \ \cdots \ h_{L-1} \end{bmatrix}^T$$

is the impulse response of the system and $w(n)$ is a zero-mean white Gaussian noise signal.

We now define the block error signal (of length $N \leq L$). For that, we assume that L is an integer multiple of N, i.e., $L = KN$. We have:

$$\mathbf{e}(m) = \mathbf{y}(m) - \hat{\mathbf{y}}(m) , \tag{8.4}$$

where m is the block time index, and

$$\mathbf{e}(m) = \begin{bmatrix} e(mN) \cdots e(mN+N-1) \end{bmatrix}^T ,$$
$$\mathbf{y}(m) = \begin{bmatrix} y(mN) \cdots y(mN+N-1) \end{bmatrix}^T$$
$$= \mathbf{X}^T(m)\mathbf{h} + \mathbf{w}(m) ,$$
$$\mathbf{X}(m) = \begin{bmatrix} \mathbf{x}(mN) \cdots \mathbf{x}(mN+N-1) \end{bmatrix} ,$$
$$\mathbf{w}(m) = \begin{bmatrix} w(mN) \cdots w(mN+N-1) \end{bmatrix}^T ,$$
$$\hat{\mathbf{y}}(m) = \begin{bmatrix} \hat{y}(mN) \cdots \hat{y}(mN+N-1) \end{bmatrix}^T$$
$$= \mathbf{X}^T(m)\hat{\mathbf{h}} .$$

It can easily be checked that $\mathbf{X}$ is a Toeplitz matrix of size $(L \times N)$. We can show that for $K = L/N$, we can write

$$\hat{\mathbf{y}}(m) = \sum_{k=0}^{K-1} \mathbf{T}(m-k)\hat{\mathbf{h}}_k , \tag{8.5}$$

where

$$\mathbf{T}(m-k) = \begin{bmatrix} x(mN-kN) & \cdots & x(mN-kN-N+1) \\ x(mN-kN+1) & \ddots & \vdots \\ \vdots & \ddots & \vdots \\ x(mN-kN+N-1) & \cdots & x(mN-kN) \end{bmatrix}$$

is an $(N \times N)$ Toeplitz matrix and

$$\hat{\mathbf{h}}_k = \begin{bmatrix} \hat{h}_{kN} \ \hat{h}_{kN+1} \cdots \hat{h}_{kN+N-1} \end{bmatrix}^T , \quad k = 0, 1, ..., K-1 ,$$

are the sub-filters of $\hat{\mathbf{h}}$. In (8.5), the filter $\hat{\mathbf{h}}$ (of length L) is partitioned into K sub-filters $\hat{\mathbf{h}}_k$ of length N and the rectangular matrix $\mathbf{X}^T$ [of size $(N \times L)$] is decomposed to K square sub-matrices of size $(N \times N)$ [136].

It is well known that a Toeplitz matrix $\mathbf{T}$ can be transformed, by doubling its size, to a circulant matrix

$$\mathbf{C} = \begin{bmatrix} \mathbf{T}' & \mathbf{T} \\ \mathbf{T} & \mathbf{T}' \end{bmatrix} ,$$

where $\mathbf{T}'$ is also a Toeplitz matrix obtained by shifting rows of $\mathbf{T}$; the diagonal of $\mathbf{C}$ is arbitrary, but it is natural and customary to set it equal to the first sample in the previous block. Using circulant matrices, the block error signal can be re-written equivalently:

$$\mathbf{e}(m) = \mathbf{y}(m) - \mathbf{W}_{N\times 2N}^{01} \hat{\mathbf{y}}'_{2N}(m) , \tag{8.6}$$

where

$$\mathbf{W}_{N\times 2N}^{01} = \begin{bmatrix} \mathbf{0}_{N\times N} \ \mathbf{I}_{N\times N} \end{bmatrix} ,$$

$$\hat{\mathbf{y}}'_{2N}(m) = \sum_{k=0}^{K-1} \mathbf{C}(m-k)\mathbf{W}^{10}_{2N\times N}\hat{\mathbf{h}}_k , \qquad (8.7)$$

$$\mathbf{C}(m-k) = \begin{bmatrix} \mathbf{T}'(m-k) & \mathbf{T}(m-k) \\ \mathbf{T}(m-k) & \mathbf{T}'(m-k) \end{bmatrix} ,$$

$$\mathbf{T}'(m-k) = \begin{bmatrix} x(mN-kN+N) & \cdots & x(mN-kN+1) \\ x(mN-kN-N+1) & \ddots & \vdots \\ \vdots & \ddots & \vdots \\ x(mN-kN-1) & \cdots & x(mN-kN+N) \end{bmatrix} ,$$

and

$$\mathbf{W}^{10}_{2N\times N} = \begin{bmatrix} \mathbf{I}_{N\times N} \\ \mathbf{0}_{N\times N} \end{bmatrix} .$$

It is also well known that a circulant matrix is easily decomposed as follows: $\mathbf{C} = \mathbf{F}^{-1}_{2N\times 2N}\mathbf{D}\mathbf{F}_{2N\times 2N}$, where $\mathbf{F}_{2N\times 2N}$ is the Fourier matrix [of size $(2N \times 2N)$] and $\mathbf{D}$ is a diagonal matrix whose elements are the discrete Fourier transform of the first column of $\mathbf{C}$. If we multiply (8.6) by $\mathbf{F}_{N\times N}$ [Fourier matrix of size $(N \times N)$], we get the error signal in the frequency domain (denoted by underbars):

$$\begin{aligned}
\underline{\mathbf{e}}(m) &= \underline{\mathbf{y}}(m) - \mathbf{G}^{01}_{N\times 2N}\underline{\hat{\mathbf{y}}}'_{2N}(m) \\
&= \underline{\mathbf{y}}(m) - \mathbf{G}^{01}_{N\times 2N} \sum_{k=0}^{K-1} \mathbf{D}(m-k)\mathbf{G}^{10}_{2N\times N}\underline{\hat{\mathbf{h}}}_k \\
&= \underline{\mathbf{y}}(m) - \mathbf{G}^{01}_{N\times 2N} \sum_{k=0}^{K-1} \mathbf{U}(m-k)\underline{\hat{\mathbf{h}}}_k \\
&= \underline{\mathbf{y}}(m) - \mathbf{G}^{01}_{N\times 2N}\underline{\mathbf{U}}(m)\underline{\hat{\mathbf{h}}} , \qquad (8.8)
\end{aligned}$$

where

$$\begin{aligned}
\underline{\mathbf{e}}(m) &= \mathbf{F}_{N\times N}\mathbf{e}(m) , \\
\underline{\mathbf{y}}(m) &= \mathbf{F}_{N\times N}\mathbf{y}(m) \\
&= \mathbf{G}^{01}_{N\times 2N}\underline{\mathbf{U}}(m)\underline{\mathbf{h}} + \underline{\mathbf{w}}(m) , \\
\underline{\mathbf{w}}(m) &= \mathbf{F}_{N\times N}\mathbf{w}(m) , \\
\mathbf{G}^{01}_{N\times 2N} &= \mathbf{F}_{N\times N}\mathbf{W}^{01}_{N\times 2N}\mathbf{F}^{-1}_{2N\times 2N} , \\
\underline{\hat{\mathbf{y}}}'_{2N}(m) &= \mathbf{F}_{2N\times 2N}\hat{\mathbf{y}}'_{2N}(m) ,
\end{aligned}$$

$$\mathbf{G}_{2N\times N}^{10} = \mathbf{F}_{2N\times 2N}\mathbf{W}_{2N\times N}^{10}\mathbf{F}_{N\times N}^{-1} ,$$

$$\underline{\hat{\mathbf{h}}}_k = \mathbf{F}_{N\times N}\hat{\mathbf{h}}_k ,$$

$$\underline{\mathbf{h}}_k = \mathbf{F}_{N\times N}\mathbf{h}_k ,$$

$$\mathbf{D}(m-k) = \mathbf{F}_{2N\times 2N}\mathbf{C}(m-k)\mathbf{F}_{2N\times 2N}^{-1} ,$$

$$\mathbf{U}(m-k) = \mathbf{D}(m-k)\mathbf{G}_{2N\times N}^{10} ,$$

$$\underline{\mathbf{D}}(m) = \begin{bmatrix} \mathbf{D}(m) \ \mathbf{D}(m-1) \ \cdots \ \mathbf{D}(m-K+1) \end{bmatrix} ,$$

$$\underline{\mathbf{U}}(m) = \begin{bmatrix} \mathbf{U}(m) \ \mathbf{U}(m-1) \ \cdots \ \mathbf{U}(m-K+1) \end{bmatrix} ,$$

$$\underline{\hat{\mathbf{h}}} = \begin{bmatrix} \underline{\hat{\mathbf{h}}}_0^T \ \underline{\hat{\mathbf{h}}}_1^T \ \cdots \ \underline{\hat{\mathbf{h}}}_{K-1}^T \end{bmatrix}^T ,$$

$$\underline{\mathbf{h}} = \begin{bmatrix} \underline{\mathbf{h}}_0^T \ \underline{\mathbf{h}}_1^T \ \cdots \ \underline{\mathbf{h}}_{K-1}^T \end{bmatrix}^T .$$

Having derived a frequency-domain error signal, we now define a frequency-domain criterion which is similar to the one proposed in [15], [14]:

$$J_{\mathrm{f}}(m) = (1-\lambda)\sum_{i=0}^{m}\lambda^{m-i}\underline{\mathbf{e}}^H(i)\underline{\mathbf{e}}(i) , \qquad (8.9)$$

where ^{H} denotes conjugate transpose and λ $(0 < \lambda < 1)$ is an exponential forgetting factor. The main advantage of using (8.9) is to take advantage of the FFT in order to have low complexity adaptive filters.

8.2.2 Normal Equation

Let ∇ be the gradient operator (with respect to $\underline{\hat{\mathbf{h}}}$). Applying the operator ∇ to the cost function J_{f}, we obtain the complex gradient vector [23]:

$$\nabla J_{\mathrm{f}}(m) = \frac{\partial J_{\mathrm{f}}(m)}{\partial \underline{\hat{\mathbf{h}}}(m)}$$

$$= -(1-\lambda)\sum_{i=0}^{m}\lambda^{m-i}\underline{\mathbf{U}}^T(i)(\mathbf{G}_{N\times 2N}^{01})^T\underline{\mathbf{y}}^*(i)$$

$$+ (1-\lambda)\left[\sum_{i=0}^{m}\lambda^{m-i}\underline{\mathbf{U}}^T(i)(\mathbf{G}_{2N\times 2N}^{01})^T\underline{\mathbf{U}}^*(i)\right]\underline{\hat{\mathbf{h}}}^*(m) , \qquad (8.10)$$

where * denotes complex conjugate,

$$\mathbf{G}_{2N\times 2N}^{01} = (\mathbf{G}_{N\times 2N}^{01})^H\mathbf{G}_{N\times 2N}^{01}$$

$$= \mathbf{F}_{2N\times 2N}\mathbf{W}_{2N\times 2N}^{01}\mathbf{F}_{2N\times 2N}^{-1} ,$$

and

$$\mathbf{W}_{2N\times 2N}^{01} = \begin{bmatrix} \mathbf{0}_{N\times N} & \mathbf{0}_{N\times N} \\ \mathbf{0}_{N\times N} & \mathbf{I}_{N\times N} \end{bmatrix} .$$

By setting the gradient of the cost function equal to zero, conjugating, noting that $(\mathbf{G}_{2N\times 2N}^{01})^H = \mathbf{G}_{2N\times 2N}^{01}$, and defining

$$\underline{\mathbf{y}}_{2N}(m) = (\mathbf{G}_{N \times 2N}^{01})^H \underline{\mathbf{y}}(m)$$

$$= \mathbf{F}_{2N \times 2N} \begin{bmatrix} \mathbf{0}_{N \times 1} \\ \mathbf{y}(m) \end{bmatrix} ,$$

we obtain the so-called *normal* equation:

$$\mathbf{S}(m)\hat{\underline{\mathbf{h}}}(m) = \mathbf{s}(m) , \tag{8.11}$$

where

$$\mathbf{S}(m) = (1 - \lambda) \sum_{i=0}^{m} \lambda^{m-i} \underline{\mathbf{U}}^H(i) \mathbf{G}_{2N \times 2N}^{01} \underline{\mathbf{U}}(i)$$

$$= \lambda \mathbf{S}(m-1) + (1 - \lambda) \underline{\mathbf{U}}^H(m) \mathbf{G}_{2N \times 2N}^{01} \underline{\mathbf{U}}(m) \tag{8.12}$$

and

$$\mathbf{s}(m) = (1 - \lambda) \sum_{i=0}^{m} \lambda^{m-i} \underline{\mathbf{U}}^H(i) \underline{\mathbf{y}}_{2N}(i)$$

$$= \lambda \mathbf{s}(m-1) + (1 - \lambda) \underline{\mathbf{U}}^H(m) \underline{\mathbf{y}}_{2N}(m)$$

$$= \lambda \mathbf{s}(m-1) + (1 - \lambda) \underline{\mathbf{U}}^H(m) (\mathbf{G}_{N \times 2N}^{01})^H \underline{\mathbf{y}}(m) . \tag{8.13}$$

If the input signal is well-conditioned (non-zero energy at all frequencies), matrix $\mathbf{S}(m)$ is nonsingular. In this case, the normal equation has a unique solution which is the optimal Wiener solution. For $N = 1$, (8.11) is strictly equivalent to the normal equation obtained by minimizing the classical recursive least-squares error criterion [4].

8.2.3 Adaptive Algorithm

There are many different ways to write the adaptive algorithm. In any case, it is derived directly from the normal equation. Enforcing this normal equation at block time indices m and $m - 1$, and using (8.12) and (8.13), we easily derive an exact adaptive algorithm:

$$\underline{\mathbf{e}}(m) = \underline{\mathbf{y}}(m) - \mathbf{G}_{N \times 2N}^{01} \underline{\mathbf{U}}(m) \hat{\underline{\mathbf{h}}}(m-1) , \tag{8.14}$$

$$\hat{\underline{\mathbf{h}}}(m) = \hat{\underline{\mathbf{h}}}(m-1) + (1 - \lambda) \mathbf{S}^{-1}(m) \underline{\mathbf{U}}^H(m) (\mathbf{G}_{N \times 2N}^{01})^H \underline{\mathbf{e}}(m) , \tag{8.15}$$

or, multiplying (8.14) by $(\mathbf{G}_{N \times 2N}^{01})^H$,

$$\underline{\mathbf{e}}_{2N}(m) = \underline{\mathbf{y}}_{2N}(m) - \mathbf{G}_{2N \times 2N}^{01} \underline{\mathbf{U}}(m) \hat{\underline{\mathbf{h}}}(m-1) , \tag{8.16}$$

$$\hat{\underline{\mathbf{h}}}(m) = \hat{\underline{\mathbf{h}}}(m-1) + (1 - \lambda) \mathbf{S}^{-1}(m) \underline{\mathbf{U}}^H(m) \underline{\mathbf{e}}_{2N}(m) , \tag{8.17}$$

where we have defined

$$\underline{\mathbf{e}}_{2N}(m) = \mathbf{F}_{2N \times 2N} \begin{bmatrix} \mathbf{0}_{N \times 1} \\ \mathbf{e}(m) \end{bmatrix} .$$

$$= (\mathbf{G}_{N \times 2N}^{01})^H \underline{\mathbf{e}}(m) .$$

Note that, for $N = 1$, this algorithm is exactly the RLS algorithm. Define the following matrix:

$$\mathbf{G}_{2L \times L}^{10} = \text{diag} \left[\mathbf{G}_{2N \times N}^{10} \cdots \mathbf{G}_{2N \times N}^{10} \right] .$$

If we multiply (8.17) by $\mathbf{G}_{2L \times L}^{10}$ and observe that $\underline{\mathbf{U}}(m) = \underline{\mathbf{D}}(m)\mathbf{G}_{2L \times L}^{10}$, we obtain the algorithm:

$$\mathbf{S}(m) = \lambda \mathbf{S}(m-1)$$
$$+ (1-\lambda)(\mathbf{G}_{2L \times L}^{10})^H \underline{\mathbf{D}}^H(m) \mathbf{G}_{2N \times 2N}^{01} \underline{\mathbf{D}}(m) \mathbf{G}_{2L \times L}^{10} , \qquad (8.18)$$

$$\underline{\mathbf{e}}_{2N}(m) = \underline{\mathbf{y}}_{2N}(m) - \mathbf{G}_{2N \times 2N}^{01} \underline{\mathbf{D}}(m)\underline{\hat{\mathbf{h}}}_{2L}(m-1) , \qquad (8.19)$$

$$\underline{\hat{\mathbf{h}}}_{2L}(m) = \underline{\hat{\mathbf{h}}}_{2L}(m-1)$$
$$+ (1-\lambda)\mathbf{G}_{2L \times L}^{10} \mathbf{S}^{-1}(m)(\mathbf{G}_{2L \times L}^{10})^H \underline{\mathbf{D}}^H(m)\underline{\mathbf{e}}_{2N}(m) , \qquad (8.20)$$

where:

$$\underline{\hat{\mathbf{h}}}_{2L}(m) = \mathbf{G}_{2L \times L}^{10} \underline{\hat{\mathbf{h}}}(m)$$
$$= \left[\underline{\hat{\mathbf{h}}}_{2N,0}^T(m) \ \underline{\hat{\mathbf{h}}}_{2N,1}^T(m) \cdots \underline{\hat{\mathbf{h}}}_{2N,K-1}^T(m) \right]^T ,$$

$$\underline{\hat{\mathbf{h}}}_{2N,k}(m) = \mathbf{F}_{2N \times 2N} \begin{bmatrix} \hat{\mathbf{h}}_k(m) \\ \mathbf{0}_{N \times 1} \end{bmatrix} .$$

The rank of the matrix $\mathbf{G}_{2L \times L}^{10}$ is equal to L. Since we have L unknowns to identify, in principle (8.20) is equivalent to (8.17). Indeed, if we multiply (8.20) by $(\mathbf{G}_{2L \times L}^{10})^H$, we obtain exactly (8.17) since $(\mathbf{G}_{2L \times L}^{10})^H \mathbf{G}_{2L \times L}^{10} = \mathbf{I}_{L \times L}$. It is interesting to see how naturally we have ended up using blocks of length $2N$ (especially for the error signal) even though we have used an error criterion with blocks of length N. We can do even better than that and rewrite the algorithm exclusively using FFTs of size $2N$. This formulation will be, by far, the most interesting because an explicit link is made with existing frequency-domain algorithms. For convenience, let us first define the $(2L \times 2L)$ matrix:

$$\mathbf{Q}(m) = (1-\lambda) \sum_{i=0}^{m} \lambda^{m-i} \underline{\mathbf{D}}^H(i) \mathbf{G}_{2N \times 2N}^{01} \underline{\mathbf{D}}(i)$$
$$= \lambda \mathbf{Q}(m-1) + (1-\lambda)\underline{\mathbf{D}}^H(m) \mathbf{G}_{2N \times 2N}^{01} \underline{\mathbf{D}}(m) . \qquad (8.21)$$

The relationship with $\mathbf{S}$ is immediate:

$$\mathbf{S}(m) = (\mathbf{G}_{2L \times L}^{10})^H \mathbf{Q}(m) \mathbf{G}_{2L \times L}^{10} . \qquad (8.22)$$

Also, define:

$$\mathbf{G}_{2L \times 2L}^{10} = \mathbf{G}_{2L \times L}^{10}(\mathbf{G}_{2L \times L}^{10})^H$$
$$= \text{diag} \left[\mathbf{G}_{2N \times 2N}^{10} \cdots \mathbf{G}_{2N \times 2N}^{10} \right] ,$$

where

$$\mathbf{G}_{2N \times 2N}^{10} = \mathbf{G}_{2N \times N}^{10}(\mathbf{G}_{2N \times N}^{10})^H$$
$$= \mathbf{F}_{2N \times 2N} \mathbf{W}_{2N \times 2N}^{10} \mathbf{F}_{2N \times 2N}^{-1} ,$$

and

$$\mathbf{W}^{10}_{2N \times 2N} = \begin{bmatrix} \mathbf{I}_{N \times N} & \mathbf{0}_{N \times N} \\ \mathbf{0}_{N \times N} & \mathbf{0}_{N \times N} \end{bmatrix} .$$

We have an interesting relation between the inverse of the two matrices $\mathbf{S}$ and $\mathbf{Q}$:

$$\mathbf{G}^{10}_{2L \times 2L} \mathbf{Q}^{-1}(m) = \mathbf{G}^{10}_{2L \times L} \mathbf{S}^{-1}(m) (\mathbf{G}^{10}_{2L \times L})^H . \tag{8.23}$$

This can be checked by post-multiplying both sides of (8.23) by $\mathbf{Q}(m) \mathbf{G}^{10}_{2L \times L}$ and noting that $\mathbf{G}^{10}_{2L \times 2L} \mathbf{G}^{10}_{2L \times L} = \mathbf{G}^{10}_{2L \times L}$.

Using (8.23), the adaptive algorithm can now be written in a much simpler and convenient way:

$$\mathbf{Q}(m) = \lambda \mathbf{Q}(m-1) + (1-\lambda) \underline{\mathbf{D}}^H(m) \mathbf{G}^{01}_{2N \times 2N} \underline{\mathbf{D}}(m) , \tag{8.24}$$

$$\underline{\mathbf{e}}_{2N}(m) = \underline{\mathbf{y}}_{2N}(m) - \mathbf{G}^{01}_{2N \times 2N} \underline{\mathbf{D}}(m) \hat{\underline{\mathbf{h}}}_{2L}(m-1) , \tag{8.25}$$

$$\hat{\underline{\mathbf{h}}}_{2L}(m) = \hat{\underline{\mathbf{h}}}_{2L}(m-1) + (1-\lambda) \mathbf{G}^{10}_{2L \times 2L} \mathbf{Q}^{-1}(m) \underline{\mathbf{D}}^H(m) \underline{\mathbf{e}}_{2N}(m) . \tag{8.26}$$

8.2.4 Convergence Analysis

In this section, we demonstrate the convergence of the algorithm in a stationary environment using (8.14) and (8.15).

Convergence in Mean

By noting that $(\mathbf{G}^{01}_{N \times 2N})^H \mathbf{G}^{01}_{N \times 2N} = \mathbf{G}^{01}_{2N \times 2N}$, (8.15) can be written as:

$$\begin{aligned}
\underline{\mathbf{h}} - \hat{\underline{\mathbf{h}}}(m) = {} & \underline{\mathbf{h}} - \hat{\underline{\mathbf{h}}}(m-1) \\
& - (1-\lambda) \mathbf{S}^{-1}(m) \underline{\mathbf{U}}^H(m) \mathbf{G}^{01}_{2N \times 2N} \underline{\mathbf{U}}(m) \left[\underline{\mathbf{h}} - \hat{\underline{\mathbf{h}}}(m-1) \right] \\
& - (1-\lambda) \mathbf{S}^{-1}(m) \underline{\mathbf{U}}^H(m) \underline{\mathbf{w}}(m) .
\end{aligned} \tag{8.27}$$

Assuming that

$$\mathbf{S}(m) \approx \mathbf{S} = E \left\{ \underline{\mathbf{U}}^H(m) \mathbf{G}^{01}_{2N \times 2N} \underline{\mathbf{U}}(m) \right\} \tag{8.28}$$

for m large, using the independence theory [66], and taking mathematical expectation of expression (8.27), we easily deduce that:

$$\begin{aligned}
E\{\underline{\boldsymbol{\varepsilon}}(m)\} &= \lambda E\{\underline{\boldsymbol{\varepsilon}}(m-1)\} \\
&= \lambda^m E\{\underline{\boldsymbol{\varepsilon}}(0)\} ,
\end{aligned} \tag{8.29}$$

where $\underline{\boldsymbol{\varepsilon}}(m) = \underline{\mathbf{h}} - \hat{\underline{\mathbf{h}}}(m)$ is the misalignment vector. Equation (8.29) says that the convergence rate of the algorithm is governed by λ. Most importantly, the rate of convergence is completely independent of the input statistics. Finally, we have:

$$\lim_{m \to \infty} E\{\underline{\boldsymbol{\varepsilon}}(m)\} = \mathbf{0}_{L \times 1} \Rightarrow \lim_{m \to \infty} E\{\hat{\mathbf{h}}(m)\} = \mathbf{h} . \tag{8.30}$$

Now, suppose that λ_t is the forgetting factor of a sample-by-sample time-domain adaptive algorithm. To have the same effective window length for the two algorithms (sample-by-sample and block-by-block), we should choose $\lambda = \lambda_t^N$. For example, the usual choice for the RLS algorithm is $\lambda_t = 1 - 1/(3L)$. In this case, a good choice for the frequency-domain algorithm is $\lambda = [1 - 1/(3L)]^N$.

Convergence in Mean Square

The convergence of the algorithm in mean is not strictly sufficient for stability [66]. An analysis of the algorithm in the mean square is required. The algorithm converges in the mean square if:

$$\lim_{m \to \infty} J_f'(m) = \text{Constant} , \tag{8.31}$$

where

$$J_f'(m) = \frac{1}{N} E\left\{ \underline{e}^H(m)\underline{e}(m) \right\} . \tag{8.32}$$

From (8.14), the error signal $\underline{e}(m)$ can be written in terms of $\underline{\varepsilon}(m)$ as:

$$\underline{e}(m) = \mathbf{G}_{N \times 2N}^{01} \underline{U}(m)\underline{\varepsilon}(m-1) + \underline{w}(m) . \tag{8.33}$$

Expression (8.32) becomes:

$$J_f'(m) = \frac{1}{N} J_{ex}(m) + \sigma_w^2 , \tag{8.34}$$

where

$$J_{ex}(m) = E\left\{ \underline{\varepsilon}^H(m-1)\underline{U}^H(m)\mathbf{G}_{2N \times 2N}^{01}\underline{U}(m)\underline{\varepsilon}(m-1) \right\} \tag{8.35}$$

is the excess mean-square error and σ_w^2 is the variance of the noise signal $w(n)$. Furthermore:

$$\begin{aligned}
J_{ex}(m) &= E\left\{ \text{tr}\left[\underline{\varepsilon}^H(m-1)\underline{U}^H(m)\mathbf{G}_{2N \times 2N}^{01}\underline{U}(m)\underline{\varepsilon}(m-1) \right] \right\} \\
&= E\left\{ \text{tr}\left[\underline{U}^H(m)\mathbf{G}_{2N \times 2N}^{01}\underline{U}(m)\underline{\varepsilon}(m-1)\underline{\varepsilon}^H(m-1) \right] \right\} \\
&= \text{tr}\left[E\left\{ \underline{U}^H(m)\mathbf{G}_{2N \times 2N}^{01}\underline{U}(m)\underline{\varepsilon}(m-1)\underline{\varepsilon}^H(m-1) \right\} \right] . \quad (8.36)
\end{aligned}$$

Invoking the independence assumption and using (8.28), we may reduce this expectation to

$$J_{ex}(m) \approx \text{tr}\left[\mathbf{SM}(m-1) \right] , \tag{8.37}$$

where

$$\mathbf{M}(m) = E\left\{ \underline{\varepsilon}(m)\underline{\varepsilon}^H(m) \right\} \tag{8.38}$$

is the misalignment correlation matrix.

Using the normal equation (8.11) and (8.13), it can easily be shown that:

$$\underline{\varepsilon}(m) = -(1 - \lambda)\mathbf{S}^{-1}(m) \sum_{i=0}^{m} \lambda^{m-i}\underline{\mathbf{U}}^{H}(i)(\mathbf{G}_{N \times 2N}^{01})^{H}\underline{\mathbf{w}}(i) \ . \tag{8.39}$$

Since we have supposed that $\mathbf{S} \approx \mathbf{S}(m)$, (8.39) can be written in a recursive way:

$$\underline{\varepsilon}(m) \approx -(1 - \lambda) \sum_{i=0}^{m} \lambda^{m-i}\mathbf{S}^{-1}\underline{\mathbf{U}}^{H}(i)(\mathbf{G}_{N \times 2N}^{01})^{H}\underline{\mathbf{w}}(i)$$

$$\approx \lambda\underline{\varepsilon}(m - 1) - (1 - \lambda)\mathbf{S}^{-1}(m)\underline{\mathbf{U}}^{H}(m)(\mathbf{G}_{N \times 2N}^{01})^{H}\underline{\mathbf{w}}(m) \ . \tag{8.40}$$

From (8.40) and (8.28), we may rewrite (8.38) as:

$$\mathbf{M}(m) \approx \frac{1 - \lambda}{2}\sigma_{w}^{2}\mathbf{S}^{-1} \ , \tag{8.41}$$

where we have used the following approximations: $\mathbf{M}(m - 1) \approx \mathbf{M}(m)$ (for m large) and $1 - \lambda^{2} \approx 2(1 - \lambda)$ (for λ close to unity). Finally:

$$J_{f}'(m) \approx \left[(1 - \lambda)\frac{L}{2N} + 1\right]\sigma_{w}^{2} \tag{8.42}$$

for m large, so that the mean-square error converges to a constant value. Moreover, the convergence of the algorithm in the mean-square error is independent of the eigenvalues of the ensemble-averaged matrix $\mathbf{S}$.

The value

$$J_{\mathrm{mis}}(m) = E\left\{\underline{\varepsilon}^{H}(m)\underline{\varepsilon}(m)\right\} \tag{8.43}$$

gives an idea of the convergence of the misalignment. Using (8.41), we deduce that

$$\begin{aligned}
J_{\mathrm{mis}}(m) &= \mathrm{tr}[\mathbf{M}(m)] \\
&= \frac{1 - \lambda}{2}\sigma_{w}^{2}\mathrm{tr}[\mathbf{S}^{-1}] \\
&= \frac{1 - \lambda}{2}\sigma_{w}^{2}\sum_{l=0}^{L-1}\frac{1}{\lambda_{s,l}} \ ,
\end{aligned} \tag{8.44}$$

where the $\lambda_{s,l}$ are the eigenvalues of the ensemble-averaged matrix $\mathbf{S}$. It is important to note the difference between the convergence of the mean-square error and the misalignment. While the mean-square error does not depend on the eigenvalues of $\mathbf{S}$, the misalignment is magnified by the inverse of the smallest eigenvalue $\lambda_{s,\min}$ of $\mathbf{S}$. The situation is worsened when the variance of the noise (σ_{w}^{2}) is large. So in practice, at some frequencies (where the signal is poorly excited) we may have a very large misalignment. In order to avoid this problem and to keep the misalignment low, the adaptive algorithm should be regularized by adding a small constant value to the diagonal of $\mathbf{S}(m)$.

8.3 Approximation and Particular Cases

We start this section by giving a very useful approximation of the algorithm proposed in Sect. 8.2. We then deduce some examples of classical efficient algorithms. This list is not exhaustive and many other algorithms can also be derived.

8.3.1 Approximation

Frequency-domain adaptive filters were first introduced to reduce the arithmetic complexity of the LMS algorithm [45]. Unfortunately, the matrix $\mathbf{Q}$ in (8.21) is not diagonal, so the proposed algorithm has a high complexity and may not be very useful in practice. Since $\mathbf{Q}$ is composed of K^2 sub-matrices, what we would like is that each of those sub-matrices be a diagonal matrix. We will argue here that $\mathbf{G}^{01}_{2N\times 2N}$ can be well approximated by $\mathbf{I}_{2N\times 2N}/2$, where $\mathbf{I}_{2N\times 2N}$ is the $2N \times 2N$ identity matrix; we then obtain the following approximate algorithm (absorbing the factor of 2 in the step size):

$$\mathbf{S}'(m) = \lambda\mathbf{S}'(m-1) + (1-\lambda)\underline{\mathbf{D}}^H(m)\underline{\mathbf{D}}(m) \ , \tag{8.45}$$

$$\underline{\mathbf{e}}_{2N}(m) = \underline{\mathbf{y}}_{2N}(m) - \mathbf{G}^{01}_{2N\times 2N}\underline{\mathbf{D}}(m)\hat{\underline{\mathbf{h}}}_{2L}(m-1) \ , \tag{8.46}$$

$$\hat{\underline{\mathbf{h}}}_{2L}(m) = \hat{\underline{\mathbf{h}}}_{2L}(m-1)$$
$$+ 2\mu(1-\lambda)\mathbf{G}^{10}_{2L\times 2L}\mathbf{S}'^{-1}(m)\underline{\mathbf{D}}^H(m)\underline{\mathbf{e}}_{2N}(m) \ , \tag{8.47}$$

where each sub-matrix of $\mathbf{S}'$ is now a diagonal matrix (as a result of this approximation, the algorithm is more attractive from a complexity point of view) and $\mu \leq 1$ is a positive number. Note that the imprecision introduced by the approximation in (8.45) will only affect the convergence rate. Obviously, we cannot permit the same kind of approximation in (8.46), because that would result in approximating a linear convolution by a circular one, which of course can have a disastrous impact in our identification problem. Now the question is the following: is the above approximation justified?

Let's examine the structure of the matrix $\mathbf{G}^{01}_{2N\times 2N}$. We have: $(\mathbf{G}^{01}_{2N\times 2N})^* = \mathbf{F}^{-1}_{2N\times 2N}\mathbf{W}^{01}_{2N\times 2N}\mathbf{F}_{2N\times 2N}$ and since $\mathbf{W}^{01}_{2N\times 2N}$ is a diagonal matrix, $(\mathbf{G}^{01}_{2N\times 2N})^*$ is a circulant matrix. Therefore, inverse transforming the diagonal of $\mathbf{W}^{01}_{2N\times 2N}$ gives the first column of $(\mathbf{G}^{01}_{2N\times 2N})^*$,

$$\mathbf{g}^* = \begin{bmatrix} g_0^* & g_1^* & \cdots & g_{2N-1}^* \end{bmatrix}^T$$
$$= \mathbf{F}^{-1}_{2N\times 2N}\begin{bmatrix} 0 & \cdots & 0 & 1 & \cdots & 1 \end{bmatrix}^T \ .$$

The elements of vector $\mathbf{g}$ can be written explicitly as:

$$g_k = \frac{1}{2N}\sum_{l=N}^{2N-1} \exp(-j2\pi kl/2N)$$
$$= \frac{(-1)^k}{2N}\sum_{l=0}^{N-1} \exp(-j\pi kl/N) \ , \tag{8.48}$$

where $j^2 = -1$. Since g_k is the sum of a geometric progression, we have:

$$g_k = \begin{cases} 0.5 , & k = 0 \\ \dfrac{(-1)^k}{2N} \dfrac{1 - \exp(-j\pi k)}{1 - \exp(-j\pi k/N)} , & k \neq 0 \end{cases}$$

$$= \begin{cases} 0.5 , & k = 0 \\ 0 , & k \text{ even} , \\ \dfrac{-1}{2N}[1 - j\cot(\dfrac{\pi k}{2N})] , & k \text{ odd} \end{cases} \tag{8.49}$$

where $N - 1$ elements of vector $\mathbf{g}$ are equal to zero. Moreover, since $(\mathbf{G}^{01}_{2N \times 2N})^H \mathbf{G}^{01}_{2N \times 2N} = \mathbf{G}^{01}_{2N \times 2N}$, then $\mathbf{g}^H \mathbf{g} = g_0 = 0.5$ and we have

$$\mathbf{g}^H \mathbf{g} - g_0^2 = \sum_{l=1}^{2N-1} |g_l|^2 = 2 \sum_{l=1}^{N-1} |g_l|^2 = \frac{1}{4} . \tag{8.50}$$

We can see from (8.50) that the first element of vector $\mathbf{g}$, i.e. g_0, is dominant, in a mean-square sense, and from (8.49) that the N first elements of $\mathbf{g}$ decrease rapidly to zero as k increases. Because of the conjugate symmetry, some of the last elements of $\mathbf{g}$ are not negligible. But, this is of little concern since $\mathbf{G}^{01}_{2N \times 2N}$ is circulant with $\mathbf{g}$ as its first column and its other columns have those non-negligible elements shifted in such a way that they are concentrated around the main diagonal. To summarize, we can say that for N large, only the very first (few) off-diagonals of $\mathbf{G}^{01}_{2N \times 2N}$ will be non-negligible while the others can be completely neglected. Thus, approximating $\mathbf{G}^{01}_{2N \times 2N}$ by a diagonal matrix, i.e. $\mathbf{G}^{01}_{2N \times 2N} \approx g_0 \mathbf{I}_{2N \times 2N} = \mathbf{I}_{2N \times 2N}/2$, is reasonable, and in this case we will have $\mu \approx 1$ for an optimal convergence rate. In the rest, we suppose that $0 < \mu \leq 1$.

8.3.2 Particular Cases

For $N = L$, $\underline{\mathbf{D}}(m) = \mathbf{D}(m)$ and $\mathbf{S}'$ is a diagonal matrix. In this case, the constrained FLMS [45] follows immediately from (8.45)-(8.47). This algorithm requires the computation of 5 FFTs of length $2L$ per block. By approximating $\mathbf{G}^{10}_{2L \times 2L}$ in (8.47) by $\mathbf{I}_{2L \times 2L}/2$, we obtain the unconstrained FLMS (UFLMS) algorithm [88] which requires only 3 FFTs per block. Many simulations (in the case $N = L$) show that the two algorithms have virtually the same performance. Table 8.1 summarizes these two algorithms.

For $N < L$, we immediately obtain the MDF algorithm by approximating (8.45) as follows:

$$\mathbf{S}'(m) \approx \text{diag}\left[\mathbf{S}_{\text{MDF}}(m) \cdots \mathbf{S}_{\text{MDF}}(m)\right] , \tag{8.51}$$

where

$$\mathbf{S}_{\text{MDF}}(m) = \lambda \mathbf{S}_{\text{MDF}}(m - 1) + (1 - \lambda)\mathbf{D}^*(m)\mathbf{D}(m) \tag{8.52}$$

is a $(2N \times 2N)$ diagonal matrix. Table 8.2 gives the constrained and unconstrained versions of this algorithm. The constrained MDF requires the

---

Table 8.1 The constrained and unconstrained FLMS algorithms

$$0 < \mu \leq 1 \text{ , adaptation step}$$

$$\lambda = [1 - 1/(3L)]^L \text{ , exponential window}$$

$$\delta \text{ , regularization factor}$$

$$\mathbf{G}^{01}_{2L \times 2L} = \mathbf{F}_{2L \times 2L} \mathbf{W}^{01}_{2L \times 2L} \mathbf{F}^{-1}_{2L \times 2L}$$

$$\mathbf{W}^{01}_{2L \times 2L} = \begin{bmatrix} \mathbf{0}_{L \times L} & \mathbf{0}_{L \times L} \\ \mathbf{0}_{L \times L} & \mathbf{I}_{L \times L} \end{bmatrix}$$

$$\mathbf{G}^{10}_{2L \times 2L} = \mathbf{F}_{2L \times 2L} \mathbf{W}^{10}_{2L \times 2L} \mathbf{F}^{-1}_{2L \times 2L}$$

$$\mathbf{W}^{10}_{2L \times 2L} = \begin{bmatrix} \mathbf{I}_{L \times L} & \mathbf{0}_{L \times L} \\ \mathbf{0}_{L \times L} & \mathbf{0}_{L \times L} \end{bmatrix}$$

$$\mathbf{S}'(m) = \lambda \mathbf{S}'(m-1) + (1-\lambda)\mathbf{D}^*(m)\mathbf{D}(m)$$

$$\underline{\mathbf{e}}_{2L}(m) = \underline{\mathbf{y}}_{2L}(m) - \mathbf{G}^{01}_{2L \times 2L}\mathbf{D}(m)\underline{\hat{\mathbf{h}}}_{2L}(m-1)$$

$$\underline{\hat{\mathbf{h}}}_{2L}(m) = \underline{\hat{\mathbf{h}}}_{2L}(m-1) + 2\mu(1-\lambda)\mathbf{G}\left[\mathbf{S}'(m) + \delta\mathbf{I}_{2L \times 2L}\right]^{-1}\mathbf{D}^*(m)\underline{\mathbf{e}}_{2L}(m)$$

$$\mathbf{G} = \mathbf{G}^{10}_{2L \times 2L} \text{ , constrained algorithm}$$

$$\mathbf{G} = \mathbf{I}_{2L \times 2L}/2 \text{ , unconstrained algorithm}$$

---

computation of $3 + 2K$ FFTs of size $2N$ per block, while the unconstrained requires only 3. Obviously, the unconstrained algorithm is much more efficient when the block size N is much smaller than the length L of the adaptive filter, but the performance (compared to the case $N = L$) is significantly degraded.

8.4 An FD Affine Projection Algorithm and a Link with the MDF

The affine projection algorithm (APA) [106] has become very popular in the last decade, especially after fast versions of this algorithm were invented [55], [125]. Its popularity comes from the fact that, for speech signals, it has almost the same performance as the FRLS, with a reduction in the number of operations of roughly a factor of 3. In this section, we derive a frequency-domain APA. A simple and elegant way to do this is to search for an algorithm of the stochastic gradient type cancelling N *a posteriori* errors [96], [100], [9]. This requirement results in an underdetermined set of linear equations of which the minimum-norm solution is chosen.

By definition, the set of N *a priori* errors $\underline{\mathbf{e}}(m)$ and N *a posteriori* errors $\underline{\mathbf{e}}_a(m)$ are:

Table 8.2 The constrained and unconstrained MDF algorithms

$$0 < \mu \leq 1 \text{ , adaptation step}$$

$$\lambda = [1 - 1/(3L)]^N \text{ , exponential window}$$

$$\delta \text{ , regularization factor}$$

$$\mathbf{G}_{2N \times 2N}^{01} = \mathbf{F}_{2N \times 2N} \mathbf{W}_{2N \times 2N}^{01} \mathbf{F}_{2N \times 2N}^{-1}$$

$$\mathbf{W}_{2N \times 2N}^{01} = \begin{bmatrix} \mathbf{0}_{N \times N} & \mathbf{0}_{N \times N} \\ \mathbf{0}_{N \times N} & \mathbf{I}_{N \times N} \end{bmatrix}$$

$$\mathbf{G}_{2N \times 2N}^{10} = \mathbf{F}_{2N \times 2N} \mathbf{W}_{2N \times 2N}^{10} \mathbf{F}_{2N \times 2N}^{-1}$$

$$\mathbf{W}_{2N \times 2N}^{10} = \begin{bmatrix} \mathbf{I}_{N \times N} & \mathbf{0}_{N \times N} \\ \mathbf{0}_{N \times N} & \mathbf{0}_{N \times N} \end{bmatrix}$$

$$\mathbf{S}_{\mathrm{MDF}}(m) = \lambda \mathbf{S}_{\mathrm{MDF}}(m - 1) + (1 - \lambda) \mathbf{D}^*(m) \mathbf{D}(m)$$

$$\underline{\mathbf{e}}_{2N}(m) = \underline{\mathbf{y}}_{2N}(m) - \mathbf{G}_{2N \times 2N}^{01} \sum_{k=0}^{K-1} \mathbf{D}(m - k) \hat{\underline{\mathbf{h}}}_{2N,k}(m - 1)$$

$$\hat{\underline{\mathbf{h}}}_{2N,k}(m) = \hat{\underline{\mathbf{h}}}_{2N,k}(m - 1)$$
$$+ 2\mu(1 - \lambda) \mathbf{G} \mathbf{D}^*(m - k) \left[\mathbf{S}_{\mathrm{MDF}}(m) + \delta \mathbf{I}_{2N \times 2N} \right]^{-1} \underline{\mathbf{e}}_{2N}(m)$$

$$k = 0, 1, ..., K - 1$$

$$\mathbf{G} = \mathbf{G}_{2N \times 2N}^{10} \text{ , constrained algorithm}$$

$$\mathbf{G} = \mathbf{I}_{2N \times 2N}/2 \text{ , unconstrained algorithm}$$

$$\underline{\mathbf{e}}(m) = \underline{\mathbf{y}}(m) - \mathbf{G}_{N \times 2N}^{01} \underline{\mathbf{D}}(m) \mathbf{G}_{2L \times L}^{10} \hat{\underline{\mathbf{h}}}(m - 1) \text{ ,} \tag{8.53}$$

$$\underline{\mathbf{e}}_{\mathrm{a}}(m) = \underline{\mathbf{y}}(m) - \mathbf{G}_{N \times 2N}^{01} \underline{\mathbf{D}}(m) \mathbf{G}_{2L \times L}^{10} \hat{\underline{\mathbf{h}}}(m) \text{ .} \tag{8.54}$$

Using these two equations plus the requirement that $\underline{\mathbf{e}}_{\mathrm{a}}(m) = \mathbf{0}_{N \times 1}$, we obtain:

$$\mathbf{G}_{N \times 2N}^{01} \underline{\mathbf{D}}(m) \mathbf{G}_{2L \times L}^{10} \left[\hat{\underline{\mathbf{h}}}(m) - \hat{\underline{\mathbf{h}}}(m - 1) \right] = \underline{\mathbf{e}}(m) \text{ .} \tag{8.55}$$

Equation (8.55) (N equations in L unknowns, $N \leq L$) is an underdetermined set of linear equations. Hence, it has an infinite number of solutions, out of which the minimum-norm solution is chosen. This results in:

$$\hat{\underline{\mathbf{h}}}(m) = \hat{\underline{\mathbf{h}}}(m - 1) + (\mathbf{G}_{2L \times L}^{10})^H \underline{\mathbf{D}}^H(m) (\mathbf{G}_{N \times 2N}^{01})^H$$
$$\times \left[\mathbf{G}_{N \times 2N}^{01} \underline{\mathbf{D}}(m) \mathbf{G}_{2L \times 2L}^{10} \underline{\mathbf{D}}^H(m) (\mathbf{G}_{N \times 2N}^{01})^H \right]^{-1} \underline{\mathbf{e}}(m) \text{ .} \tag{8.56}$$

Multiplying (8.56) by $\mathbf{G}_{2L \times L}^{10}$, we obtain:

$$\hat{\underline{\mathbf{h}}}_{2L}(m) = \hat{\underline{\mathbf{h}}}_{2L}(m - 1) + \mathbf{G}_{2L \times 2L}^{10} \underline{\mathbf{D}}^H(m) (\mathbf{G}_{N \times 2N}^{01})^H$$
$$\times \left[\mathbf{G}_{N \times 2N}^{01} \underline{\mathbf{D}}(m) \mathbf{G}_{2L \times 2L}^{10} \underline{\mathbf{D}}^H(m) (\mathbf{G}_{N \times 2N}^{01})^H \right]^{-1} \underline{\mathbf{e}}(m) \text{ ,} \tag{8.57}$$

or, equivalently, each sub-filter is updated as follows:

$$\hat{\underline{h}}_{2N,k}(m) = \hat{\underline{h}}_{2N,k}(m-1) + \mathbf{G}_{2N \times 2N}^{10} \mathbf{D}^*(m-k)(\mathbf{G}_{N \times 2N}^{01})^H$$
$$\times \left[\mathbf{G}_{N \times 2N}^{01} \underline{\mathbf{D}}(m) \mathbf{G}_{2L \times 2L}^{10} \underline{\mathbf{D}}^H(m)(\mathbf{G}_{N \times 2N}^{01})^H \right]^{-1} \underline{e}(m) , \quad (8.58)$$
$$k = 0, 1, ..., K - 1 .$$

Furthermore, it can be easily checked that:

$$(\mathbf{G}_{N \times 2N}^{01})^H \left[\mathbf{G}_{N \times 2N}^{01} \underline{\mathbf{D}}(m) \mathbf{G}_{2L \times 2L}^{10} \underline{\mathbf{D}}^H(m)(\mathbf{G}_{N \times 2N}^{01})^H \right]^{-1}$$
$$= \left[\underline{\mathbf{D}}(m) \mathbf{G}_{2L \times 2L}^{10} \underline{\mathbf{D}}^H(m) \right]^{-1} (\mathbf{G}_{N \times 2N}^{01})^H , \quad (8.59)$$

so that (8.58) is simplified to

$$\hat{\underline{h}}_{2N,k}(m) = \hat{\underline{h}}_{2N,k}(m-1) + \mathbf{G}_{2N \times 2N}^{10} \mathbf{D}^*(m-k)$$
$$\times \left[\underline{\mathbf{D}}(m) \mathbf{G}_{2L \times 2L}^{10} \underline{\mathbf{D}}^H(m) \right]^{-1} \underline{e}_{2N}(m) , \quad (8.60)$$
$$k = 0, 1, ..., K - 1 ,$$

which is also an MDF-type algorithm except that the cross-power spectrum matrix is estimated with a rectangular window instead of an exponential window as in Sect. 8.3.2.

8.5 Generalization to the Multichannel Case

The generalization to the multichannel case is rather straightforward. Therefore, in this section we only highlight some important steps and state the algorithms. For convenience, we will use the same notation as previously employed. Let P be the number of channels. Our definition of multichannel is that we have a system with P input signals x_p, $p = 1, 2, ..., P$, and one output signal y (see Fig. 8.2). By definition, the error signal at time n is:

$$e(n) = y(n) - \sum_{p=1}^{P} \hat{\mathbf{h}}_p^T \mathbf{x}_p(n) , \quad (8.61)$$

where $\hat{\mathbf{h}}_p$ is the estimated impulse response of the pth channel,

$$\hat{\mathbf{h}}_p = \left[\hat{h}_{p,0} \; \hat{h}_{p,1} \; \cdots \; \hat{h}_{p,L-1} \right]^T$$

and

$$\mathbf{x}_p(n) = \left[x_p(n) \; x_p(n-1) \; \cdots \; x_p(n-L+1) \right]^T .$$

Now the block error signal is defined as:

$$\mathbf{e}(m) = \mathbf{y}(m) - \sum_{p=1}^{P} \mathbf{X}_p^T(m) \hat{\mathbf{h}}_p , \quad (8.62)$$

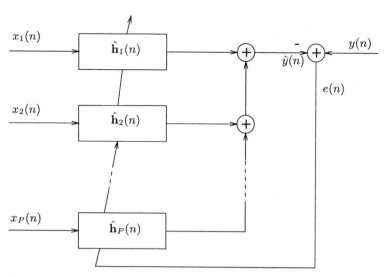

Fig. 8.2. Multichannel adaptive filtering

where $\mathbf{e}$ and $\mathbf{y}$ are vectors of e and y respectively, and all matrices $\mathbf{X}_p$ are Toeplitz of size $(L \times N)$. In the frequency-domain, we have:

$$
\begin{aligned}
\underline{\mathbf{e}}(m) &= \underline{\mathbf{y}}(m) - \mathbf{G}^{01}_{N\times 2N} \sum_{p=1}^{P} \sum_{k=0}^{K-1} \mathbf{D}_p(m-k) \mathbf{G}^{10}_{2N\times N} \underline{\hat{\mathbf{h}}}_{p,k} \\
&= \underline{\mathbf{y}}(m) - \mathbf{G}^{01}_{N\times 2N} \sum_{p=1}^{P} \sum_{k=0}^{K-1} \mathbf{U}_p(m-k) \underline{\hat{\mathbf{h}}}_{p,k} \\
&= \underline{\mathbf{y}}(m) - \mathbf{G}^{01}_{N\times 2N} \sum_{p=1}^{P} \underline{\mathbf{U}}_p(m) \underline{\hat{\mathbf{h}}}_p \\
&= \underline{\mathbf{y}}(m) - \mathbf{G}^{01}_{N\times 2N} \underline{\mathbf{U}}(m) \underline{\hat{\mathbf{h}}} \,,
\end{aligned}
\tag{8.63}
$$

where

$$\hat{\mathbf{h}}_{p,k} = \left[\hat{h}_{p,kN} \; \hat{h}_{p,kN+1} \; \cdots \; \hat{h}_{p,kN+N-1} \right]^T ,$$

$$\underline{\hat{\mathbf{h}}}_{p,k} = \mathbf{F}_{N \times N} \hat{\mathbf{h}}_{p,k} ,$$

$$\underline{\hat{\mathbf{h}}}_p = \left[\underline{\hat{\mathbf{h}}}_{p,0}^T \; \underline{\hat{\mathbf{h}}}_{p,1}^T \; \cdots \; \underline{\hat{\mathbf{h}}}_{p,K-1}^T \right]^T ,$$

$$\underline{\hat{\mathbf{h}}} = \left[\underline{\hat{\mathbf{h}}}_1^T \; \underline{\hat{\mathbf{h}}}_2^T \; \cdots \; \underline{\hat{\mathbf{h}}}_P^T \right]^T ,$$

$$\mathbf{D}_p(m - k) = \mathbf{F}_{2N \times 2N} \mathbf{C}_p(m - k) \mathbf{F}_{2N \times 2N}^{-1} ,$$

$$\mathbf{U}_p(m - k) = \mathbf{D}_p(m - k) \mathbf{G}_{2N \times N}^{10} ,$$

$$\underline{\mathbf{U}}_p(m) = \left[\mathbf{U}_p(m) \; \mathbf{U}_p(m - 1) \; \cdots \; \mathbf{U}_p(m - K + 1) \right] ,$$

$$\underline{\mathbf{U}}(m) = \left[\underline{\mathbf{U}}_1(m) \; \underline{\mathbf{U}}_2(m) \; \cdots \; \underline{\mathbf{U}}_P(m) \right] ,$$

$$\underline{\mathbf{D}}_p(m) = \left[\mathbf{D}_p(m) \; \mathbf{D}_p(m - 1) \; \cdots \; \mathbf{D}_p(m - K + 1) \right] ,$$

$$\underline{\mathbf{D}}(m) = \left[\underline{\mathbf{D}}_1(m) \; \underline{\mathbf{D}}_2(m) \; \cdots \; \underline{\mathbf{D}}_P(m) \right] ,$$

$$\mathbf{G}_{2PL \times PL}^{10} = \mathrm{diag} \left[\mathbf{G}_{2N \times N}^{10} \; \cdots \; \mathbf{G}_{2N \times N}^{10} \right] ,$$

$$\underline{\mathbf{U}}(m) = \underline{\mathbf{D}}(m) \mathbf{G}_{2PL \times PL}^{10} .$$

The size of the matrix $\underline{\mathbf{U}}$ is $(2N \times PL)$ and the length of $\underline{\hat{\mathbf{h}}}$ is PL.

Minimizing the criterion defined in (8.9), we obtain the normal equation for the multichannel case:

$$\mathbf{S}(m) \underline{\hat{\mathbf{h}}}(m) = \mathbf{s}(m) , \tag{8.64}$$

where

$$\mathbf{S}(m) = \lambda \mathbf{S}(m - 1)$$
$$+ (1 - \lambda)(\mathbf{G}_{2PL \times PL}^{10})^H \underline{\mathbf{D}}^H(m) \mathbf{G}_{2N \times 2N}^{01} \underline{\mathbf{D}}(m) \mathbf{G}_{2PL \times PL}^{10} \tag{8.65}$$

is a $(PL \times PL)$ matrix and

$$\mathbf{s}(m) = \lambda \mathbf{s}(m - 1) + (1 - \lambda)(\mathbf{G}_{2PL \times PL}^{10})^H \underline{\mathbf{D}}^H(m) \underline{\mathbf{y}}_{2N}(m) \tag{8.66}$$

is a $(PL \times 1)$ vector. Assuming that the P input signals are not perfectly pairwise coherent, the normal equation has a unique solution which is the optimal Wiener solution.

Define the following variables:

$$\mathbf{G}_{2PL \times 2PL}^{10} = \mathrm{diag} \left[\mathbf{G}_{2N \times 2N}^{10} \; \cdots \; \mathbf{G}_{2N \times 2N}^{10} \right]$$

and

$$\underline{\hat{\mathbf{h}}}_{2PL}(m) = \mathbf{G}_{2PL \times PL}^{10} \underline{\hat{\mathbf{h}}}(m) .$$

Using the same approach and definitions as in Sect. 8.2, we get the multichannel frequency-domain adaptive algorithm:

$$\mathbf{Q}(m) = \lambda\mathbf{Q}(m-1) + (1-\lambda)\underline{\mathbf{D}}^H(m)\mathbf{G}_{2N\times 2N}^{01}\underline{\mathbf{D}}(m) , \tag{8.67}$$

$$\underline{\mathbf{e}}_{2N}(m) = \underline{\mathbf{y}}_{2N}(m) - \mathbf{G}_{2N\times 2N}^{01}\underline{\mathbf{D}}(m)\hat{\underline{\mathbf{h}}}_{2PL}(m-1) , \tag{8.68}$$

$$\hat{\underline{\mathbf{h}}}_{2PL}(m) = \hat{\underline{\mathbf{h}}}_{2PL}(m-1)$$
$$+ (1-\lambda)\mathbf{G}_{2PL\times 2PL}^{10}\mathbf{Q}^{-1}(m)\underline{\mathbf{D}}^H(m)\underline{\mathbf{e}}_{2N}(m) . \tag{8.69}$$

Now, to have a more efficient algorithm, we use the approximation given in Sect. 8.3. Finally, we have:

$$\mathbf{S}'(m) = \lambda\mathbf{S}'(m-1) + (1-\lambda)\underline{\mathbf{D}}^H(m)\underline{\mathbf{D}}(m) , \tag{8.70}$$

$$\underline{\mathbf{e}}_{2N}(m) = \underline{\mathbf{y}}_{2N}(m) - \mathbf{G}_{2N\times 2N}^{01}\underline{\mathbf{D}}(m)\hat{\underline{\mathbf{h}}}_{2PL}(m-1) , \tag{8.71}$$

$$\hat{\underline{\mathbf{h}}}_{2PL}(m) = \hat{\underline{\mathbf{h}}}_{2PL}(m-1)$$
$$+ 2\mu(1-\lambda)\mathbf{G}_{2PL\times 2PL}^{10}\mathbf{S}'^{-1}(m)\underline{\mathbf{D}}^H(m)\underline{\mathbf{e}}_{2N}(m) . \tag{8.72}$$

As an example, we give in Table 8.3 the two-channel FLMS ($N = L$) [15] which happens to be very useful in the context of stereophonic acoustic echo cancellation [14], [40]. The two-channel MDF is described in detail in [11], [12].

8.6 Conclusions

In many applications when an adaptive filter is required, frequency-domain algorithms (when well optimized) can be extremely good alternatives to time-domain algorithms or adaptive algorithms in subbands. First, the complexity can be made low by utilizing the computational efficiency of the FFT. The delay can be kept as small as desired since the block size is independent of the length of the adaptive filter. Finally, the convergence rate can be relatively good if some parameters of these algorithms such as the exponential window, regularization factor, and adaptation step size are properly chosen. The objective of this chapter was to present a general framework for frequency-domain adaptive filtering. We have shown that an exact algorithm can be derived from the normal equation after minimizing a block least-squares criterion in the frequency domain. We have shown the convergence conditions for this algorithm and have introduced various approximations that lead to well-known algorithms such as the FLMS, UFLMS, and MDF. Other approximations are possible that may lead to some interesting algorithms. We have also shown that APA and MDF are strongly related. Finally, we have generalized some of these ideas to the multichannel case.

Table 8.3 The constrained and unconstrained two-channel FLMS algorithms

$$0 < \mu \le 1 \text{ , adaptation step}$$

$$\lambda = [1 - 1/(3L)]^L \text{ , exponential window}$$

$$\delta_1 \text{ , } \delta_2 \text{ , regularization factors}$$

$$\mathbf{G}^{01}_{2L \times 2L} = \mathbf{F}_{2L \times 2L} \mathbf{W}^{01}_{2L \times 2L} \mathbf{F}^{-1}_{2L \times 2L}$$

$$\mathbf{W}^{01}_{2L \times 2L} = \begin{bmatrix} \mathbf{0}_{L \times L} & \mathbf{0}_{L \times L} \\ \mathbf{0}_{L \times L} & \mathbf{I}_{L \times L} \end{bmatrix}$$

$$\mathbf{G}^{10}_{2L \times 2L} = \mathbf{F}_{2L \times 2L} \mathbf{W}^{10}_{2L \times 2L} \mathbf{F}^{-1}_{2L \times 2L}$$

$$\mathbf{W}^{10}_{2L \times 2L} = \begin{bmatrix} \mathbf{I}_{L \times L} & \mathbf{0}_{L \times L} \\ \mathbf{0}_{L \times L} & \mathbf{0}_{L \times L} \end{bmatrix}$$

$$\mathbf{S}'_{p,q}(m) = \lambda \mathbf{S}'_{p,q}(m-1) + (1-\lambda)\mathbf{D}^*_p(m)\mathbf{D}_q(m), \ p,q = 1,2$$

$$\tilde{\mathbf{S}}'_{p,p}(m) = \mathbf{S}'_{p,p}(m) + \delta_p \mathbf{I}_{2L \times 2L}, \ p = 1,2$$

$$\mathbf{S}'_p(m) = \tilde{\mathbf{S}}'_{p,p}(m) \left[\mathbf{I}_{2L \times 2L} - \underline{\boldsymbol{\Gamma}}^H(m)\underline{\boldsymbol{\Gamma}}(m) \right], \ p = 1,2$$

$$\underline{\boldsymbol{\Gamma}}(m) = \left[\tilde{\mathbf{S}}'_{1,1}(m)\tilde{\mathbf{S}}'_{2,2}(m) \right]^{-1/2} \mathbf{S}'_{1,2}(m), \text{ coherence matrix}$$

$$\underline{\mathbf{e}}_{2L}(m) = \underline{\mathbf{y}}_{2L}(m) - \mathbf{G}^{01}_{2L \times 2L} \left[\mathbf{D}_1(m)\hat{\underline{\mathbf{h}}}_{2L,1}(m-1) + \mathbf{D}_2(m)\hat{\underline{\mathbf{h}}}_{2L,2}(m-1) \right]$$

$$\hat{\underline{\mathbf{h}}}_{2L,1}(m) = \hat{\underline{\mathbf{h}}}_{2L,1}(m-1)$$
$$+ 2\mu(1-\lambda)\mathbf{G}\mathbf{S}'^{-1}_1(m) \left[\mathbf{D}^*_1(m) - \mathbf{S}'_{1,2}(m)\tilde{\mathbf{S}}'^{-1}_{2,2}(m)\mathbf{D}^*_2(m) \right] \underline{\mathbf{e}}_{2L}(m)$$

$$\hat{\underline{\mathbf{h}}}_{2L,2}(m) = \hat{\underline{\mathbf{h}}}_{2L,2}(m-1)$$
$$+ 2\mu(1-\lambda)\mathbf{G}\mathbf{S}'^{-1}_2(m) \left[\mathbf{D}^*_2(m) - \mathbf{S}'_{2,1}(m)\tilde{\mathbf{S}}'^{-1}_{1,1}(m)\mathbf{D}^*_1(m) \right] \underline{\mathbf{e}}_{2L}(m)$$

$$\mathbf{G} = \mathbf{G}^{10}_{2L \times 2L} \text{ , constrained algorithm}$$

$$\mathbf{G} = \mathbf{I}_{2L \times 2L}/2 \text{ , unconstrained algorithm}$$

9. Multichannel Acoustic Echo and Double-Talk Handling: A Frequency-Domain Approach

9.1 Introduction

A multichannel frequency-domain adaptive algorithm was presented in Chap. 8 (see also [15]). The multichannel frequency-domain algorithm has been shown to work very well in the two-channel acoustic echo cancellation application [40]. It has a fairly low computational complexity compared to the fast recursive least-squares algorithm (FRLS) [40]. Furthermore, it is an inherently stable algorithm, well suited for a fixed-point implementation. Our objective in this chapter is to provide a complete solution, based on the multichannel frequency-domain adaptive algorithm, which handles both echo cancellation and double-talk.

Double-talk, i.e., when the speech of the two talkers arrives simultaneously at the canceler, is best handled by a double-talk detector (DTD) combined with built-in robustness in the adaptive algorithm, as presented in Chaps. 2, 3, and 6. A frequency-domain equivalent to the fast normalized cross-correlation DTD [16] (Chap. 6) as well as a robust version of the multichannel frequency-domain adaptive algorithm are proposed in this chapter.

The chapter is organized as follows. Section 9.2 introduces the multichannel echo cancellation problem, definitions, and notation needed in the description of the adaptive algorithm and double-talk detector. A robust version of the frequency-domain (FD) algorithm presented in Chap. 8 is derived in Sect. 9.3. Section 9.4 reviews the test statistic of the normalized cross-correlation DTD and presents a frequency-domain version based on a pseudo-coherence estimate. A method for detecting echo path changes is discussed in Sect. 9.5. The proposed frequency-domain system is evaluated in Sect. 9.6, where we study double-talk handling and tracking of echo path changes. Section 9.7 summarizes the results and presents some conclusions.

9.2 Definitions and Notation for the Multichannel FD Algorithm

This section gives the definitions and notation of the multichannel frequency-domain adaptive algorithm presented in Chap. 8 for a block size equal to the

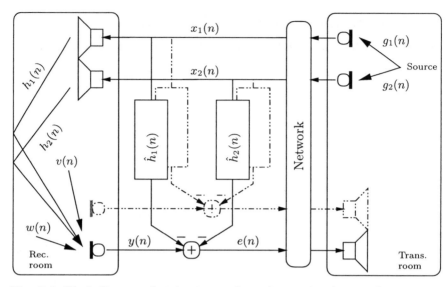

Fig. 9.1. Block diagram of a generic two-channel acoustic echo canceler

length of the adaptive filter. Figure 9.1 shows the basic block diagram of a two-channel acoustic echo canceler. (In a real-life situation, we need an echo canceler for the "transmission room" as well.) Communication is hands-free between a transmission room and a receiving room where we denote the signals picked up by the microphones in the transmission room by $x_1(n)$, $x_2(n)$, and the return signal picked up by one of the microphones in the receiving room by $y(n)$. The receiving room signal is in general composed of echo, ambient noise $w(n)$, and possibly receiving room speech $v(n)$, which in this scenario is referred to as double-talk. Thus we have the receiving room signal model:

$$y(n) = y_e(n) + v(n) + w(n) , \tag{9.1}$$

where $y_e(n) = \sum_{p=1}^{2} h_p * x_p(n)$ is the echo ($*$ denotes convolution).

For the general P-channel case, we assume that we have P incoming transmission room signals, $x_p(n)$, $p = 1, \ldots, P$. Define the P input excitation vectors as

$$\mathbf{x}_p(n) = \begin{bmatrix} x_p(n) \, x_p(n-1) \cdots x_p(n-L+1) \end{bmatrix}^T , \ p = 1, 2, ..., P ,$$

where the superscript ^{T} denotes transposition.

The error signal at time n between one arbitrary microphone output $y(n)$ in the receiving room and its estimate is

$$e(n) = y(n) - \sum_{p=1}^{P} \hat{y}_p(n) = y(n) - \sum_{p=1}^{P} \hat{\mathbf{h}}_p^T \mathbf{x}_p(n) , \tag{9.2}$$

where

$$\hat{\mathbf{h}}_p = \begin{bmatrix} \hat{h}_{p,0} \ \hat{h}_{p,1} \cdots \hat{h}_{p,L-1} \end{bmatrix}^T , \ p = 1, 2, ..., P ,$$

are the P modeling filters. In general, we have P microphones in the receiving room as well, i.e. P return channels, and thus P^2 adaptive filters. From now on, though, we will just look at one return channel because the derivation is similar for the other $P - 1$ channels.

In order to present the frequency-domain adaptive algorithm, we need to define the following block signals,

$$\mathbf{e}(m) = [e(mL) \ \cdots \ e(mL + L - 1)]^T , \tag{9.3}$$
$$\mathbf{y}(m) = [y(mL) \ \cdots \ y(mL + L - 1)]^T , \tag{9.4}$$

where the block length is chosen to be equal to the length of the adaptive filter L. We can now write the estimated return signal block as

$$\hat{\mathbf{y}}_p(m) = [\mathbf{x}_p(mL) \ \cdots \ \mathbf{x}_p(mL + L - 1)]^T \hat{\mathbf{h}}_p , \tag{9.5}$$
$$= \mathbf{X}_p^T(m)\hat{\mathbf{h}}_p , \ p = 1, \ldots, P ,$$

where $\mathbf{X}_p$ is an $(L \times L)$ Toeplitz matrix. The block error can be written as,

$$\mathbf{e}(m) = \mathbf{y}(m) - \sum_{p=1}^{P} \mathbf{X}_p^T(m)\hat{\mathbf{h}}_p . \tag{9.6}$$

We would like to compute the Toeplitz-vector product in (9.6) with fewer multiplications than a general matrix-vector product. This can be achieved by embedding the Toeplitz matrix in a circulant matrix [61]. A circulant matrix $\mathbf{C}_p$, $(2L \times 2L)$, can be formed from $\mathbf{X}_p$ as

$$\mathbf{C}_p(m) = \begin{bmatrix} \mathbf{X}_p'(m) \ \mathbf{X}_p(m) \\ \mathbf{X}_p(m) \ \mathbf{X}_p'(m) \end{bmatrix}^T , \tag{9.7}$$

where $\mathbf{X}_p'(m)$ is also a Toeplitz matrix obtained by shifting rows of $\mathbf{X}_p^T(m)$; the diagonal of $\mathbf{C}_p$ is arbitrary, but it is natural and customary to set it to the first sample in the previous sequence $\mathbf{x}_p(mL - L)$, i.e. $x_p(mL - L)$. We can now rewrite the block error equation, (9.6), as

$$\begin{bmatrix} \mathbf{0}_{L\times 1} \\ \mathbf{e}(m) \end{bmatrix} = \begin{bmatrix} \mathbf{0}_{L\times 1} \\ \mathbf{y}(m) \end{bmatrix} - \mathbf{W}^{01} \sum_{p=1}^{P} \hat{\mathbf{y}}_p'(m) , \tag{9.8}$$

where

$$\mathbf{W}^{01} = \begin{bmatrix} \mathbf{0}_{L\times L} \ \mathbf{0}_{L\times L} \\ \mathbf{0}_{L\times L} \ \mathbf{I}_{L\times L} \end{bmatrix} , \tag{9.9}$$

$$\hat{\mathbf{y}}_p'(m) = \mathbf{C}_p(m) \begin{bmatrix} \hat{\mathbf{h}}_p \\ \mathbf{0}_{L\times 1} \end{bmatrix} . \tag{9.10}$$

The fundamental property that makes it possible to reduce the computational complexity of the frequency-domain algorithms is that $\mathbf{C}_p$ can be diagonalized by the Fourier matrix $\mathbf{F}$, i.e.

$$\mathbf{C}_p = \mathbf{F}^{-1}\mathbf{D}_p\mathbf{F} , \tag{9.11}$$

where $\mathbf{F}$ is the Fourier matrix,

$$\mathbf{F} = \begin{bmatrix} 1 & 1 & \cdots & 1 \\ 1 & e^{-j\frac{2\pi}{2L}} & \cdots & e^{-j\frac{2\pi(2L-1)}{2L}} \\ \vdots & \vdots & \ddots & \vdots \\ 1 & e^{-j\frac{2\pi(2L-1)}{2L}} & \cdots & e^{-j\frac{2\pi(2L-1)^2}{2L}} \end{bmatrix} , \tag{9.12}$$

and $\mathbf{D}_p$ is a diagonal matrix whose elements are the discrete Fourier transform of the first column of $\mathbf{C}_p$.

Transformation of the above derived equations into the frequency-domain (denoted by underbars) gives

$$\underline{\mathbf{e}}(m) = \mathbf{F}\begin{bmatrix} \mathbf{0}_{L\times 1} \\ \mathbf{e}(m) \end{bmatrix}$$

$$= \underline{\mathbf{y}}(m) - \sum_{p=1}^{P}\mathbf{G}^{01}\underline{\hat{\mathbf{y}}}'_p(m)$$

$$= \underline{\mathbf{y}}(m) - \mathbf{G}^{01}\sum_{p=1}^{P}\mathbf{D}_p(m)\underline{\hat{\mathbf{h}}}_p$$

$$= \underline{\mathbf{y}}(m) - \mathbf{G}^{01}\mathbf{D}(m)\underline{\hat{\mathbf{h}}} , \tag{9.13}$$

where

$$\underline{\mathbf{y}}(m) = \mathbf{F}\begin{bmatrix} \mathbf{0}_{L\times 1} \\ \mathbf{y}(m) \end{bmatrix} , \tag{9.14}$$

$$\underline{\hat{\mathbf{y}}}'_p(m) = \mathbf{F}\hat{\mathbf{y}}'_p(m) , \tag{9.15}$$

$$\mathbf{G}^{01} = \mathbf{F}\mathbf{W}^{01}\mathbf{F}^{-1} , \tag{9.16}$$

$$\underline{\hat{\mathbf{h}}}_p = \mathbf{F}\begin{bmatrix} \hat{\mathbf{h}}_p \\ \mathbf{0}_{L\times 1} \end{bmatrix} , \tag{9.17}$$

$$\underline{\hat{\mathbf{h}}} = \begin{bmatrix} \underline{\hat{\mathbf{h}}}_1^T & \underline{\hat{\mathbf{h}}}_2^T & \cdots & \underline{\hat{\mathbf{h}}}_P^T \end{bmatrix}^T , \tag{9.18}$$

$$\mathbf{D}(m) = [\mathbf{D}_1(m) \ \mathbf{D}_2(m) \ \cdots \ \mathbf{D}_P(m)] . \tag{9.19}$$

With this formalism, a robust multichannel frequency-domain adaptive algorithm can be derived by minimizing a non-quadratic criterion with respect to $\underline{\hat{\mathbf{h}}}$. The resulting algorithm is robust against large disturbances resulting from double-talk detection errors. We would also like to point out that the derivation of the robust version of the multichannel frequency-domain algorithm here is somewhat simplified from the general derivation of the non-robust version in Sect. 8.2. Here, we will only consider the unconstrained algorithm and assume that the block length N equals the filter length L. The justification for using the unconstrained algorithm is that room impulse responses decay exponentially, so that their tails do not cause serious wraparound problems.

9.3 An Outlier-Robust Multichannel FD Algorithm

As in derivations of time-domain adaptive algorithms, we start by forming a criterion that is minimized with respect to its filter coefficients. Most commonly, the choice is a quadratic criterion that corresponds to a maximum likelihood estimator when the underlying noise distribution is Gaussian. However, in our case, we choose to look at the maximum likelihood criterion derived from a non-Gaussian noise assumption. Modeling the noise with a probability density function (PDF) with a tail that is heavier than the Gaussian PDF gives us a non-quadratic function to minimize, which results in an outlier-robust algorithm [68]. We choose to work with the following criterion:

$$J(\hat{\underline{h}}) = \sum_{n=mL}^{mL+L-1} \rho\left[\frac{|e(n)|}{s}\right], \tag{9.20}$$

where $\rho[\cdot]$ is a convex function and s is a real positive scale factor for block m, as discussed below. The nonlinear function $\rho[\cdot]$ can be chosen in many ways; however, as long as it is chosen to have a bounded derivative, the resulting algorithm inherits robust properties [68]. Robust methods have been proposed and developed for a variety of time-domain echo cancellation algorithms in [51] and Chaps. 2, 3, and 6. In this chapter, we extend this concept to the frequency-domain case. A variant of this frequency-domain approach was proposed in [48]. A good choice of $\rho[\cdot]$ is [68]

$$\rho(|z|) = \begin{cases} \dfrac{|z|^2}{2}, & |z| \le k_0 \\ k_0|z| - \dfrac{k_0^2}{2}, & |z| \ge k_0 \end{cases}, \tag{9.21}$$

where k_0 is a constant controlling the robustness of the algorithm.
Adaptive Newton-type algorithms [75] minimize the criterion (9.20) by using the recursion:

$$\hat{\underline{h}}(m) = \hat{\underline{h}}(m-1) - \mu' \mathbf{S}_{\psi'}^{-1} \nabla J\left[\hat{\underline{h}}(m-1)\right], \tag{9.22}$$

where $\nabla J\left(\hat{\underline{h}}\right) = \partial/\partial\hat{\underline{h}}^* J$ is the gradient of the optimization criterion w.r.t. $\hat{\underline{h}}$, $\mathbf{S}_{\psi'}$ is an approximation of the expected value of the Hessian $\nabla^2 J = \partial/\partial\hat{\underline{h}}^* (\nabla J)^H$, and μ' is the relaxation parameter.

In order to proceed, we need to calculate the gradient and Hessian of (9.20). First, we write one complex conjugated element of the error vector $\mathbf{e}(m)$ as

$$
\begin{aligned}
e^*(n) &= \left[\mathbf{0}_{L\times 1}^H \ \mathbf{e}^H(m)\right] \mathbf{1}_{n-mL+L} \\
&= \underline{\mathbf{e}}^H(m)\mathbf{F}^{-H}\mathbf{1}_{n-mL+L} \\
&= \left[\mathbf{y}^H(m) - \hat{\underline{\mathbf{h}}}^H \mathbf{D}^H(m)(\mathbf{G}^{01})^H\right] \mathbf{F}^{-H}\mathbf{1}_{n-mL+L} , \\
& \qquad n = mL, \ldots , mL + L - 1 ,
\end{aligned}
\tag{9.23}
$$

where $\mathbf{1}_n$ is a $2L \times 1$ vector containing a 1 in position n and zeros in all other positions. Then, the gradient is found by the following:

$$
\begin{aligned}
\nabla J &= \frac{\partial}{\partial \hat{\underline{\mathbf{h}}}^*} J = \sum_{n=mL}^{mL+L-1} \frac{\partial}{\partial \hat{\underline{\mathbf{h}}}^*} \rho \left[\frac{|e(n)|}{s}\right] \\
&= \sum_{n=mL}^{mL+L-1} \frac{\partial}{\partial \hat{\underline{\mathbf{h}}}^*} [e^*(n)] \frac{\partial}{\partial e(n)} |e(n)| \frac{\partial}{\partial |e(n)|} \rho \left[\frac{|e(n)|}{s}\right] \\
&= \sum_{n=mL}^{mL+L-1} \frac{\partial}{\partial \hat{\underline{\mathbf{h}}}^*} [e^*(n)] \rho' \left[\frac{|e(n)|}{s}\right] \frac{\text{sign}\,[e(n)]}{s} \\
&= - \sum_{n=mL}^{mL+L-1} \mathbf{D}^H(m)(\mathbf{G}^{01})^H\mathbf{F}^{-H}\mathbf{1}_{n-mL+L}\psi \left[\frac{|e(n)|}{s}\right] \frac{\text{sign}\,[e(n)]}{s} \\
&= -\frac{1}{s} \mathbf{D}^H(m)(\mathbf{G}^{01})^H\mathbf{F}^{-H}\psi\,[\mathbf{e}(m)] \\
&= -\frac{1}{2L\,s} \mathbf{D}^H(m)\mathbf{F}\psi\,[\mathbf{e}(m)] ,
\end{aligned}
\tag{9.24}
$$

where vector $\psi\,[\mathbf{e}(m)]$ is defined as

$$
\psi\,[\mathbf{e}(m)] =
\begin{bmatrix}
\mathbf{0}_{L\times 1} \\[4pt]
\psi \left[\dfrac{|e(mL)|}{s}\right] \text{sign}\,[e(mL)] \\[4pt]
\vdots \\[4pt]
\psi \left[\dfrac{|e(mL + L - 1)|}{s}\right] \text{sign}\,[e(mL + L - 1)]
\end{bmatrix} ,
\tag{9.25}
$$

and

$$
\psi\,(|z|) = \rho'\,(|z|) = \min\{|z|, k_0\} .
\tag{9.26}
$$

Combining (9.22) and (9.24), we find the robust frequency-domain algorithm:

$$
\hat{\underline{\mathbf{h}}}(m) = \hat{\underline{\mathbf{h}}}(m - 1) + \frac{\mu'}{2L\,s}\mathbf{S}_{\psi'}^{-1}(m)\,\mathbf{D}^H(m)\mathbf{F}\psi\,[\mathbf{e}(m)] .
\tag{9.27}
$$

We now derive the Hessian and estimate its expected value $\mathbf{S}_{\psi'}$. The Hessian is expressed as

$$\nabla^2 J = \frac{\partial}{\partial \hat{\mathbf{h}}^*} (\nabla J)^H = -\frac{1}{s} \frac{\partial}{\partial \hat{\mathbf{h}}^*} \psi^H [\mathbf{e}(m)] \mathbf{F}^{-1} \mathbf{G}^{01} \mathbf{D}(m) . \tag{9.28}$$

If we look at one non-zero element of $\psi^H [\mathbf{e}(m)]$,

$$\frac{\partial}{\partial \hat{\mathbf{h}}^*} \left(\psi \left[\frac{|e(n)|}{s} \right] \mathrm{sign} [e(n)] \right)$$

$$= \frac{\partial}{\partial \hat{\mathbf{h}}^*} [e^*(n)] \frac{\partial}{\partial e(n)} |e(n)| \frac{\partial}{\partial |e(n)|} \psi \left[\frac{|e(n)|}{s} \right] \mathrm{sign} [e(n)]$$

$$= - \mathbf{D}^H(m) \mathbf{G}^{01} \mathbf{F}^{-H} \mathbf{1}_{n-mL+L} \frac{\mathrm{sign} [e(n)]}{s} \psi' \left[\frac{|e(n)|}{s} \right] \mathrm{sign} [e(n)] , \tag{9.29}$$

we can write

$$\frac{\partial}{\partial \hat{\mathbf{h}}^*} \psi^H [\mathbf{e}(m)] = -\frac{1}{s} \mathbf{D}^H(m) (\mathbf{G}^{01})^H \mathbf{F}^{-H}$$

$$\times \begin{bmatrix} \mathbf{0}_{L \times L} & 0 & \cdots & & 0 \\ \mathbf{0}_{1 \times L} \, \psi' \left[\frac{|e(mL)|}{s} \right] & 0 & \cdots & & 0 \\ \vdots & 0 & \ddots & & \vdots \\ \vdots & \vdots & \cdots & & 0 \\ \mathbf{0}_{1 \times L} & 0 & \cdots & 0 & \psi' \left[\frac{|e(mL+L-1)|}{s} \right] \end{bmatrix}$$

$$= -\frac{1}{s} \mathbf{D}^H(m) (\mathbf{G}^{01})^H \mathbf{F}^{-H} \mathbf{\Psi}' [\mathbf{e}(m)] , \tag{9.30}$$

where ψ' is the derivative of ψ and

$$\mathbf{\Psi}' [\mathbf{e}(m)] = \mathrm{diag} \left\{ \mathbf{0}_{1 \times L} \, \psi' \left[\frac{e(mL)}{s} \right] \quad \cdots \quad \psi' \left[\frac{e(mL+L-1)}{s} \right] \right\} . \tag{9.31}$$

Combining (9.28) and (9.30) gives

$$\nabla^2 J = \frac{1}{s^2} \mathbf{D}^H(m) (\mathbf{G}^{01})^H \mathbf{F}^{-H} \mathbf{\Psi}' [\mathbf{e}(m)] \mathbf{F}^{-1} \mathbf{G}^{01} \mathbf{D}(m) . \tag{9.32}$$

Finally, by recursive averaging of (9.32), we find an estimate of its expected value,

$$\mathbf{S}_{\psi'}(m) = \frac{(1 - \lambda)}{s^2} \sum_{l=0}^{m} \lambda^{m-l} \mathbf{D}^H(l) \mathbf{G}^{01} \mathbf{F}^{-H} \mathbf{\Psi}' [\mathbf{e}(l)] \mathbf{F}^{-1} \mathbf{G}^{01} \mathbf{D}(l) , \tag{9.33}$$

where λ $(0 < \lambda < 1)$ is an exponential forgetting factor.

The matrix $\mathbf{S}_{\psi'}$ is in general non-diagonal. Hence (9.27) is of high computational complexity. To simplify this algorithm we would like to use the same approximation as for the non-robust frequency-domain method described in Sect. 8.3. We start by approximating $\mathbf{\Psi}'[\mathbf{e}(m)]$ as

$$\mathbf{\Psi}'[\mathbf{e}(m)] = \psi'_{\min}(m)\mathbf{W}^{01} , \tag{9.34}$$

where

$$\psi'_{\min}(m) = \min_{0 \le l \le L-1} \left\{ \psi' \left[\frac{|e(mL+l)|}{s} \right] \right\} . \tag{9.35}$$

Since $\mathbf{F}^{-H}\mathbf{W}^{01}\mathbf{F}^{-1} = \mathbf{G}^{01}/(2L)$ and $(\mathbf{G}^{01})^2 = \mathbf{G}^{01}$, (9.33) is now approximated as

$$\mathbf{S}_{\psi'}(m) = \frac{(1-\lambda)}{s^2} \sum_{l=0}^{m} \lambda^{m-l} \frac{\psi'_{\min}(m)}{2L} \mathbf{D}^H(l)\mathbf{G}^{01}\mathbf{D}(l)$$

$$= \lambda\mathbf{S}_{\psi'}(m-1) + (1-\lambda)\frac{\psi'_{\min}(m)}{s^2 2L}\mathbf{D}^H(m)\mathbf{G}^{01}\mathbf{D}(m) . \tag{9.36}$$

Further approximating $\mathbf{G}^{01}$ in (9.36) with $\mathbf{I}/2$ according to Sect. 8.3 and assuming that $\psi'_{\min}(m)$ varies slowly, we finally get a simplified version of (9.27) as:

$$\hat{\underline{\mathbf{h}}}(m) = \hat{\underline{\mathbf{h}}}(m-1) + \frac{2\,\mu's}{\psi'_{\min}(m)}\mathbf{S}'^{-1}(m)\,\mathbf{D}^H(m)\mathbf{F}\psi[\mathbf{e}(m)] , \tag{9.37}$$

where

$$\mathbf{S}'(m) = \lambda\mathbf{S}'(m-1) + (1-\lambda)\,\mathbf{D}^H(m)\mathbf{D}(m) . \tag{9.38}$$

Note that (9.38) is the same recursive relation as (8.45). Thus, the main new feature here is the robust error function $s\psi[\mathbf{e}(m)]/\psi'_{\min}(m)$. We have normalized the relaxation parameter μ' as in Chap. 8 by letting $\mu' = \mu(1-\lambda)$ with $0 < \mu \le 1$. The value of the derivative $\psi'_{\min}(m)$ is either 1 or 0. Obviously, it is not appropriate to let $\psi'_{\min}(m)$ become zero since it is a divisor in (9.37). We therefore bound it in such a way that (9.37) always remains stable, i.e, $\psi'_{\min}(m) \ge \mu$. Thus, we have

$$\psi'_{\min}(m) = \max\left[\mu, \min_{0 \le l \le L-1}\left\{\psi'\left[\frac{|e(mL+l)|}{s}\right]\right\}\right] . \tag{9.39}$$

This gives freedom for control of the algorithm during convergence and tracking, as discussed later in Sect. 9.5. Finally, we note that in the non-robust case, $\psi[\mathbf{e}(m)] = \mathbf{e}(m)$ and, with $s/\psi'_{\min}(m) = 1$, (9.37) reduces to an unconstrained version of (8.47), as it should.

9.3.1 Scale Factor Estimation

An important part of the algorithm is the estimation of the scale factor s which has been discussed in Chaps. 2, 3, and 6. Here, we chose the scale factor estimate as

$$s(m+1) = \lambda_s s(m) + (1 - \lambda_s)\frac{s(m)}{L\beta} \sum_{n=mL}^{mL+L-1} \psi\left[\frac{|e(n)|}{s(m)}\right], \qquad (9.40)$$

$$s(0) = \sigma_x .$$

The constant β is related to k_0 in (9.26) and chosen such that the scale factor is an unbiased estimate of the standard deviation of the ambient noise when the noise is assumed to be Gaussian (see Chap. 2 for more details).

In general, the robust algorithm tolerates more double-talk detection errors than a non-robust algorithm. This simplifies the design and makes parameter choices of the DTD less critical. In the simulations section, we show the performance of the non-robust as well as the robust frequency-domain algorithm.

9.3.2 A Regularized Two-Channel Version

The simulations in this chapter use a two-channel algorithm. We therefore present this version in more detail.

It is known that adaptive filters may have large misalignment for frequency bands with poor excitation. In particular, poor excitation will produce small eigenvalues of the matrix $\mathbf{S}'(m)$ in (9.38), thus increasing the algorithm's noise sensitivity. It is therefore important to regularize this algorithm carefully such that the update step is small for sensitive bands. Regularization is performed according to,

$$\tilde{\mathbf{S}}(m) = \mathbf{S}'(m) + \text{diag}\{\delta_{1,0}\ldots\delta_{1,2L-1}\ \delta_{2,0}\ldots\delta_{2,2L-1}\} , \qquad (9.41)$$

where $\delta_{p,l}$, $p = 1, 2$, $l = 0,\ldots,2L-1$, are chosen as a fraction of the input signal power. This gives us the two-channel algorithm:

$$\mathbf{e}(m) = \mathbf{y}(m) - \mathbf{W}^{01}\mathbf{F}^{-1}\left[\mathbf{D}_1(m)\hat{\underline{\mathbf{h}}}_1(m-1) + \mathbf{D}_2(m)\hat{\underline{\mathbf{h}}}_2(m-1)\right] , \qquad (9.42)$$

$$\hat{\underline{\mathbf{h}}}_1(m) = \hat{\underline{\mathbf{h}}}_1(m-1) + \frac{2\mu's}{\psi'_{\min}(m)}\mathbf{S}_1^{-1}(m) \times$$
$$\left[\mathbf{D}_1^*(m) - \tilde{\mathbf{S}}_{1,2}(m)\tilde{\mathbf{S}}_{2,2}^{-1}(m)\,\mathbf{D}_2^*(m)\right]\mathbf{F}\psi\left[\mathbf{e}(m)\right] , \qquad (9.43)$$

$$\hat{\underline{\mathbf{h}}}_2(m) = \hat{\underline{\mathbf{h}}}_2(m-1) + \frac{2\mu's}{\psi'_{\min}(m)}\mathbf{S}_2^{-1}(m) \times$$
$$\left[\mathbf{D}_2^*(m) - \tilde{\mathbf{S}}_{2,1}(m)\tilde{\mathbf{S}}_{1,1}^{-1}(m)\,\mathbf{D}_1^*(m)\right]\mathbf{F}\psi\left[\mathbf{e}(m)\right] , \qquad (9.44)$$

where $\tilde{\mathbf{S}}_{p,q}(m)$ are diagonal sub-matrices of the regularized matrix $\tilde{\mathbf{S}}(m)$,

$$\mathbf{S}_p(m) = \tilde{\mathbf{S}}_{p,p}(m)\left[\mathbf{I}_{2L\times 2L} - \mathbf{\Gamma}^H(m)\mathbf{\Gamma}(m)\right] , \qquad (9.45)$$

and $\mathbf{\Gamma}(m)$ is the diagonal coherence matrix:

$$\mathbf{\Gamma}(m) = \left[\tilde{\mathbf{S}}_{1,1}(m)\tilde{\mathbf{S}}_{2,2}(m)\right]^{-1/2}\tilde{\mathbf{S}}_{1,2}(m) . \tag{9.46}$$

Finally, it should be mentioned that this algorithm has been presented with no overlap of the data. Overlapping, i.e. updating the filter coefficients more often than every L samples, can be easily done by computing the FFTs on overlapped data. Overlapping is described by a parameter o which means that the coefficients are updated every L/o samples. Most often, one chooses $o = 1$, 2, 4, or 8. The advantage of updating more often than once per block is that the algorithm converges and tracks echo path changes faster.

9.4 Double-Talk Detection Based on a Pseudo-Coherence Measure

Double-talk detectors (DTDs), of any kind, essentially operate in the same manner. A decision variable, ξ, is formed from available signals, $x(n)$, $y(n)$, $e(n)$. This variable is compared to a preset threshold T. A decision, whether double-talk is present or not, is made according to

$$\xi < T, \Rightarrow \text{ double-talk} ,$$
$$\xi \geq T, \Rightarrow \text{ no double-talk} .$$

See Chap. 6 for more details on double-talk detectors.

A double-talk detection statistic was proposed in [16], [28] that has the potential of fulfilling the properties of an "optimum" statistic (see Chap. 6). This variable is defined in terms of the normalized cross-correlation vector between the excitation vector $\mathbf{x}(n)$ and the return signal $y(n)$. The multichannel normalized cross-correlation detection statistic [16] (Chap. 6) is defined as

$$\xi = \sqrt{\mathbf{r}^T(\sigma_y^2\mathbf{R})^{-1}\mathbf{r}} = ||\mathbf{c}_{xy}||_2 , \tag{9.47}$$

where the variables are defined as:

$$\mathbf{R} = E\{\mathbf{x}(n)\mathbf{x}^T(n)\} , \tag{9.48}$$
$$\mathbf{r} = E\{\mathbf{x}(n)y(n)\} , \tag{9.49}$$
$$\sigma_y^2 = E\{y^2(n)\} , \tag{9.50}$$
$$\mathbf{x}(n) = \left[\mathbf{x}_1^T(n)\ \mathbf{x}_2^T(n)\cdots\mathbf{x}_P^T(n)\right]^T , \tag{9.51}$$

and $\mathbf{c}_{xy}$ is the normalized cross-correlation vector between $\mathbf{x}$ and y,

$$\mathbf{c}_{xy} = (\sigma_y^2\mathbf{R})^{-1/2}\mathbf{r} . \tag{9.52}$$

Moreover, we can rewrite ξ as:

$$\xi = \frac{\sqrt{\mathbf{h}^T\mathbf{R}\mathbf{h}}}{\sqrt{\mathbf{h}^T\mathbf{R}\mathbf{h} + \sigma_v^2}} , \tag{9.53}$$

where $\mathbf{h} = \begin{bmatrix} \mathbf{h}_1^T & \mathbf{h}_2^T & \dots & \mathbf{h}_P^T \end{bmatrix}^T$ is the (stacked) vector of impulse responses and σ_v^2 is the variance of speech in the receiving room. It is easily deduced from this form that for $v(n) = 0$, $\xi = 1$ and for $v(n) \neq 0$, $\xi < 1$. Note also that ξ is independent of the echo path when $v(n) = 0$. Hence, our threshold, T, should be chosen in the range, $0 < T < 1$; typically, $0.85 \leq T \leq 0.99$. In Chap. 6, it was shown that the normalized cross-correlation (NCC) detector is in fact estimating the echo path as is done in the echo canceler. The concept of an NCC DTD and an adaptive echo canceler can therefore be regarded as a background and a foreground filter respectively, which is similar to the two-path principle for double-talk detection [104] (see Chap. 6).

Evaluation of the NCC DTD shows that it has very high performance, i.e. high probability of detection while maintaining a low probability of false alarm, even when echo path changes occur. The objective of this section is therefore to derive and present a corresponding frequency-domain version of the NCC DTD, that can be used together with the frequency-domain adaptive algorithm presented in the previous section.

Calculation of (9.47) in the frequency-domain can be made from what we call the pseudo-coherence (PC) vector,

$$\mathbf{c}_{xy}^{\mathrm{pc}} = (2L^2 \sigma_y^2 \mathbf{S})^{-1/2} \mathbf{s} \,, \tag{9.54}$$

where

$$\mathbf{S} = E\{\mathbf{D}^H(m)\mathbf{G}^{01}\mathbf{D}(m)\} \,, \tag{9.55}$$

$$\mathbf{s} = E\{\mathbf{D}^H(m)\underline{\mathbf{y}}(m)\} \,, \tag{9.56}$$

$\mathbf{D}(m)$ is defined in (9.19), $\mathbf{G}^{01}$ is defined in (9.16), and $\underline{\mathbf{y}}(m)$ is defined in (9.14). Looking at (9.54), we see that each cross-spectrum bin of (9.56) is normalized by the corresponding spectrum of the input signal, $\mathbf{x}$. What differentiates (9.54) from being the true coherence is that it is not normalized by the corresponding spectrum of the output signal y but by the total power of the output signal, σ_y^2; hence we call it *pseudo coherence*. A detection statistic can be defined in the frequency-domain using (9.54),

$$\xi^{\mathrm{pc}} = ||\mathbf{c}_{xy}^{\mathrm{pc}}||_2 \,. \tag{9.57}$$

A practical double-talk detection statistic can now be realized by using estimated quantities in (9.54) and slightly rewriting the numerator,

$$\xi^2(m) = \frac{\mathbf{s}^H(m)\mathbf{S}^{-1}(m)\mathbf{s}(m)}{\sigma_{\underline{\mathbf{y}}}^2(m)} = \frac{\mathbf{s}^H(m)\hat{\underline{\mathbf{h}}}_{\mathrm{b}}(m)}{\sigma_{\underline{\mathbf{y}}}^2(m)} \,, \tag{9.58}$$

where we squared the statistic and dropped the superscript ^{pc} for simplicity, and have defined

$$\hat{\underline{\mathbf{h}}}_{\mathrm{b}}(m) = \mathbf{S}^{-1}(m)\mathbf{s}(m) \tag{9.59}$$

as an equivalent "background" filter, and the estimated quantities are given by

$$s(m) = \lambda_b s(m-1) + (1 - \lambda_b) \mathbf{D}^H(m) \underline{\mathbf{y}}(m) , \tag{9.60}$$

$$\mathbf{S}(m) = \lambda_b \mathbf{S}(m-1) + (1 - \lambda_b) \mathbf{D}^H(m) \mathbf{G}^{01} \mathbf{D}(m) , \tag{9.61}$$

$$\sigma_{\underline{\mathbf{y}}}^2(m) = \lambda_b \sigma_{\underline{\mathbf{y}}}^2(m-1) + (1 - \lambda_b) \underline{\mathbf{y}}^H(m) \underline{\mathbf{y}}(m) . \tag{9.62}$$

As a further simplification, the background echo path estimate $\hat{\underline{\mathbf{h}}}_b(m)$ in (9.58) is obtained by using a separate filter for the DTD [not to be confused with the foreground estimate (9.37)]. This background filter is estimated as

$$\hat{\underline{\mathbf{h}}}_b(m) = \hat{\underline{\mathbf{h}}}_b(m-1) + (1 - \lambda_b) \, \mathbf{S}'^{-1}(m) \, \mathbf{D}^H(m) \underline{\mathbf{e}}_b(m) , \tag{9.63}$$

where $\mathbf{S}'(m)$ is defined in (9.38) and

$$\underline{\mathbf{e}}_b(m) = \underline{\mathbf{y}}(m) - \mathbf{G}^{01} \mathbf{D}(m) \hat{\underline{\mathbf{h}}}_b(m-1) . \tag{9.64}$$

The background estimate should be adapted with a smaller forgetting factor, λ_b, than that of the foreground filter, λ. By this choice, we ensure that the DTD detects double-talk fast and alerts the foreground filter before it diverges.

Finally, we can show that (9.58) and the square of (9.47) are equivalent by looking at the inner product of the cross-spectrum vector (9.56) and the frequency-domain echo path vector, as follows. We know that $\underline{\mathbf{y}} = \mathbf{G}^{01} \mathbf{D}(m) \hat{\underline{\mathbf{h}}}$ in the noiseless case and $\mathbf{G}^{01} = (\mathbf{G}^{01})^H \mathbf{G}^{01}$. Therefore, noting that

$$\hat{\mathbf{h}}_p^H \mathbf{F}^{-H} E\{\mathbf{C}_p(m) \mathbf{W}_{2L \times 2L}^{01} \mathbf{C}_q(m)\} \mathbf{F} \hat{\mathbf{h}}_q$$
$$= \hat{\mathbf{h}}_p^T E\{\mathbf{X}_p(m) \mathbf{X}_q^T(m)\} \hat{\mathbf{h}}_q , \tag{9.65}$$

we have

$$\hat{\underline{\mathbf{h}}}^H s = \hat{\underline{\mathbf{h}}}^H E\{\mathbf{D}^H(m)(\mathbf{G}^{01})^H \mathbf{G}^{01} \mathbf{D}(m)\} \hat{\underline{\mathbf{h}}}$$
$$= 2L \hat{\underline{\mathbf{h}}}^H \mathbf{F}^{-H} E\{\mathbf{C}^H(m) \mathbf{W}^{01} \mathbf{C}(m)\} \mathbf{F}^{-1} \hat{\underline{\mathbf{h}}}$$
$$= 2L^2 \hat{\mathbf{h}}^T \mathbf{R} \hat{\mathbf{h}}$$
$$= 2L^2 \mathbf{r}^T \mathbf{R}^{-1} \mathbf{r} , \tag{9.66}$$

where

$$\mathbf{C}(m) = [\mathbf{C}_1(m) \ \mathbf{C}_2(m) \ \dots \ \mathbf{C}_P(m)] , \tag{9.67}$$

$\mathbf{C}_p(m)$ is defined in (9.7), and $\mathbf{W}^{01}$ is defined in (9.9). A similar type of calculation shows that $E\{\sigma_{\underline{\mathbf{y}}}^2\} = E\{\underline{\mathbf{y}}^H(m) \underline{\mathbf{y}}(m)\} = 2L^2 \sigma_y^2$.

9.4.1 Choosing an Appropriate Detection Threshold

A rough idea of how to choose the threshold T in the DTD can be found from a normalized misalignment formula for the adaptive algorithm[1]. We therefore start by deriving this formula for the (non-robust) frequency-domain

---

[1] The formulae in this section are derived for the single-channel case. However, we can obtain the two-channel formulae by assuming that both transmission signals have equal variance and replacing L by $2L$.

algorithms as well as the RLS algorithms (see Chap. 6). Assume that the input signal as well as ambient noise and near-end signal are white. This is of course a crude approximation for speech. However, the resulting formula fairly accurately estimates the normalized MSE for input signals and noise with arbitrary spectral shape. Let the total background signal be defined as the sum of double-talk and ambient noise,

$$b(n) = v(n) + w(n) \ . \tag{9.68}$$

Then, assuming independence, we have

$$\sigma_b^2 = \sigma_v^2 + \sigma_w^2 \ . \tag{9.69}$$

The adaptive weight misalignment is defined as

$$\boldsymbol{\varepsilon}(m) = \mathbf{h} - \hat{\mathbf{h}}(m) \ . \tag{9.70}$$

From Chap. 8, we know that the expected misalignment energy is given by

$$E\{\|\boldsymbol{\varepsilon}\|^2\} = 2L \, E\{\|\boldsymbol{\epsilon}\|^2\} = \mathrm{tr}\left\{\frac{(1-\lambda)}{1+\lambda}\sigma_b^2\mathbf{S}^{-1}\right\} = \frac{(1-\lambda)}{1+\lambda}\frac{L\sigma_b^2}{N\sigma_x^2} \ , \tag{9.71}$$

since we have assumed white noise input. Here, either $N = L/o$ for the case when the data is overlapped by a factor o, as discussed in Sect. 9.3.2, or N is the block length as defined in Sect. 8.2, depending on which algorithm we analyze.

Define the echo-to-background ratio:

$$\mathrm{EBR} = \frac{\|\mathbf{h}\|^2\sigma_x^2}{\sigma_b^2} = \frac{\|\mathbf{h}\|^2\sigma_x^2}{\sigma_v^2 + \sigma_w^2} \ . \tag{9.72}$$

The normalized misalignment formula is found by normalizing (9.71) and using (9.72):

$$\frac{E\{\|\boldsymbol{\varepsilon}(m)\|^2\}}{\|\mathbf{h}\|^2} = \frac{1-\lambda}{1+\lambda}\frac{L}{N\,\mathrm{EBR}} \ . \tag{9.73}$$

This formula is valid for the frequency-domain algorithm as well as the time-domain RLS algorithm where $N = 1$ and usually $L \gg 1$. In this case for $\lambda \approx 1$, (9.73) becomes (6.1). According to Chap. 8, we should choose

$$\lambda = \left(1 - \frac{1}{K_0L}\right)^N \ , \tag{9.74}$$

where K_0 is the "forgetting" constant in the algorithms. Rewriting (9.73), we get

$$
\frac{E\{\|\boldsymbol{\varepsilon}(m)\|^2\}}{\|\mathbf{h}\|^2} = \frac{1 - \left(1 - \frac{1}{K_0 L}\right)^N}{1 + \left(1 - \frac{1}{K_0 L}\right)^N} \frac{L}{N \, \mathrm{EBR}}
$$

$$
\approx \frac{1}{\left(2K_0 - \frac{N}{L}\right) \mathrm{EBR}} . \tag{9.75}
$$

Define the echo-to-ambient-noise ratio

$$
\mathrm{ENR} = \frac{\|\mathbf{h}\|^2 \sigma_x^2}{\sigma_w^2} . \tag{9.76}
$$

For the PC (as well as the FNCC, $N = 1$) DTD, the detection statistic is then given by,

$$
\xi^2 = \frac{\mathbf{h}^T \mathbf{R} \mathbf{h}}{\mathbf{h}^T \mathbf{R} \mathbf{h} + \sigma_w^2} = \frac{\mathrm{ENR}}{1 + \mathrm{ENR}} \tag{9.77}
$$

when double-talk is not present, and

$$
\xi^2 = \frac{\mathbf{h}^T \mathbf{R} \mathbf{h}}{\mathbf{h}^T \mathbf{R} \mathbf{h} + \sigma_w^2 + \sigma_v^2} = \frac{\mathrm{EBR}}{1 + \mathrm{EBR}} \tag{9.78}
$$

when double-talk is present. Using (9.77), clearly we should choose the maximum value of T according to

$$
T < \sqrt{\frac{\mathrm{ENR}}{1 + \mathrm{ENR}}} \tag{9.79}
$$

in order to avoid declaring double-talk because of background noise. Using (9.75) and (9.78) we find that the minimum value of T should be chosen according to

$$
T \geq \sqrt{\frac{1}{\dfrac{E\{\|\boldsymbol{\varepsilon}(m)\|^2\}}{\|\mathbf{h}\|^2}(2K_0 - N/L) + 1}} \tag{9.80}
$$

to ensure the desired performance during double-talk. Finally, the approximate average performance during double-talk for an arbitrary threshold T is lower bounded by:

$$
\frac{E\{\|\boldsymbol{\varepsilon}(m)\|^2\}}{\|\mathbf{h}\|^2} \geq \frac{1/T^2 - 1}{2K_0 - N/L} . \tag{9.81}
$$

9.5 A Tracking Improvement Technique

The convergence rate of the algorithm is important when there is an echo path change. The pseudo-coherence-based DTD in Sect. 9.4 and the foreground adaptive filter in Sect. 9.3 give us the possibility to detect when an echo path change has occurred. Since the DTD is a background adaptive filter and the echo canceler is a foreground adaptive filter, both estimating the same echo path, we can look at the performance of each filter and conclude when the convergence rate of the foreground filter should increase.

The background filter is always updated regardless of double-talk. Assume that double-talk has not been declared. We can then deduce the following: If the background residual error power is smaller than the foreground error power, and at the same time, the robustness function ψ is in its saturated region, the probability that an echo path change has occurred is high and we can decide to let $\psi'_{\min}(m)$ in (9.39) reach its minimum value (μ), thereby increasing the convergence rate.

9.6 Evaluation and Simulations

In this section, we evaluate the performance of the pseudo-coherence-based DTD by estimating its receiver operating characteristic (ROC). The performance of the whole system, double-talk detector and echo canceler (both non-robust and robust), is shown when echo path changes occur in the receiving room as well as in the transmission room and when the system is exposed to double-talk. All results are directly applicable to the general multichannel case.

9.6.1 Receiver Operating Characteristic

When evaluating a detector, it is customary to show the probability of detection, P_d, or equivalently, probability of miss, $P_m = 1 - P_d$, versus the probability of false alarm, P_f. This has been previously suggested in [28], [52] and was used to evaluate the fast NCC DTD in Chap. 6. One problem that has to be kept in mind, though, is that speech has a time-varying power level. This may give less consistent results when estimating P_d and P_f than what would be the case for a stationary signal. Nevertheless, for comparative studies of DTDs, the estimated ROC is still a useful performance metric. In this section, we only evaluate the ROC for the single-channel version of the DTD because of lengthy simulations; the two-channel version will have performance comparable to that of the single-channel version.

Estimation of P_f and P_d using a set of speech data obtained from a database was made according to the procedure presented in [28]. The data we use contain the same sentences as in Chap. 6, originating from three male and two female talkers. Simulation details are as follows:

Echo path. The echo path used in this section originates from the same path as used in Sect. 6.6.1 of Chap. 6. In this section though, it is not down-sampled but truncated to a total length of 2048 coefficients, which

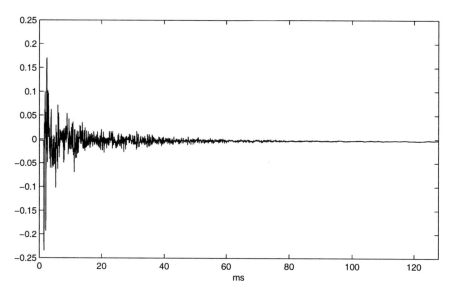

Fig. 9.2. Echo path response used to simulate the acoustic path when the ROC of the PC DTD is estimated

corresponds to a length of 128 ms at a 16 kHz sampling rate (see Fig. 9.2). It is also normalized so that $\sigma_{y_e}^2 = \sigma_x^2$ for the actual speech data in order to maintain a constant average power ratio between the echo and the receiving room speech.

Probability of false alarm. When estimating the probability of false alarm, we use sentences that correspond to a speech sequence of 21.8 seconds at a 16-kHz sampling rate. The echo-to-ambient-noise ratio, ENR $= \sigma_{y_e}^2/\sigma_w^2$, is set to 1000 (30 dB), which is typical of many office environments. **Probability of miss.** The probability of a miss is estimated using 5 seconds of transmission room speech from one of the male talkers. As receiving room speech, 8 sentences each about 2 seconds long are used. In our case we investigate the performance when the average echo-to-background ratio, EBR $= \sigma_{y_e}^2/(\sigma_v^2 + \sigma_w^2)$, is 1 (0 dB). This is natural in a desktop situation where we can assume equally strong talkers and distances between microphone and loudspeaker/talker.

The tested thresholds of the PC detector are chosen such that its probability of false alarm goes from approximately 0 to about 0.12. For these thresholds, we then estimate the corresponding probability of miss (see [28] for details). Results from the simulations are shown as the ROC in Fig. 9.3. If we compare this frequency-domain result with the ROC for the time-domain FNCC DTD in Chap. 6, we find that these detectors have the similar performance.

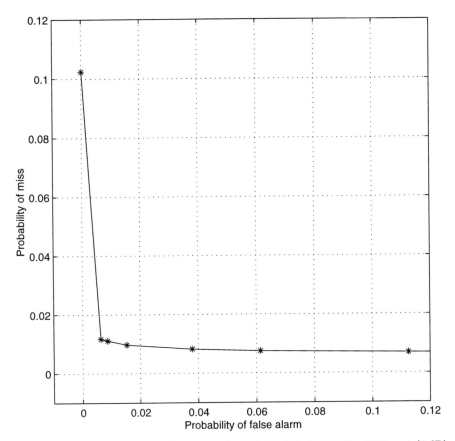

Fig. 9.3. Receiver operating characteristic (ROC) of the PC DTD. EBR = 1 (0 dB) and ENR = 1000 (30 dB)

9.6.2 Data for the Two-Channel Case

For the two-channel simulations, we use real-life speech data described in [41]. The source data are the same as used in the previous section. These two-channel recordings, however, were made in the HuMaNet room B [21]. When double-talk behavior and changes in the receiving room are studied, the transmission room signals have been pre-processed with a nonlinearity so that the echo canceler converges to a stable unique solution (see [19], Chap. 5).

Recorded receiving room data are used. In these situations, the microphone and loudspeaker setup in the receiving room is shown in Fig. 6.8, and estimated acoustic responses and magnitude functions are shown in Figs. 6.9 and 6.10. The sampling rate is 16 kHz. Data and general algorithm settings are:

Transmission Room Speech: Stereo recordings with the same male talker as in the previous section. The transmission room speech is pre-processed before emitted in the receiving room according to Chap. 5 [19]:

$$x_1 = x_1 + \frac{\alpha}{2}\left[x_1 + |x_1|\right] , \tag{9.82}$$

$$x_2 = x_2 + \frac{\alpha}{2}\left[x_2 - |x_2|\right] , \tag{9.83}$$

where the subscript 1 or 2 denotes either left or right channel, respectively. Thus, the positive half-wave is added to the left channel and the negative to the right. The distortion parameter of the nonlinearity is α.

Ambient noise level: ENR $= \sigma_{y_e}^2/\sigma_w^2 = 6000$ (38 dB).

Adaptive filter parameters: $L = 1024$ (64 ms), $\lambda = [1 - 1/(3L)]^{L/o}$, $o = 4$, $\mu = 0.5$, $\underline{\mathbf{h}}(0) = \mathbf{0}$, $\varrho = 0.99$ (relaxation parameter, see Table 9.1).

DTD parameters: $\lambda_b = [1 - 2/(3L)]^{L/o}$, ENR $= 1000$ (30 dB),
- **Non-robust adaptive filter:** $T = 0.99 \Rightarrow P_m \approx 0.012$, $P_f \approx 0.06$.
- **Robust adaptive filter:** $T = 0.92 \Rightarrow P_m \approx 0.102$, $P_f \approx 0.0$.

Table 9.1 presents the modified system of two-channel PC DTD and frequency-domain algorithm used in the simulations of this section. Although not shown in the table, we also incorporated the tracking logic discussed in Sect. 9.5.

The performance is measured by means of the normalized mean-square error (MSE), which is estimated as

$$\text{MSE}(n) = \frac{\text{LPF}([y_e(n) - \hat{y}(n)]^2)}{\text{LPF}([y_e(n)]^2)} , \tag{9.84}$$

where LPF denotes a first-order lowpass filter with a pole at 0.999 (time constant $= 1/16$ s at 16 kHz sampling rate).

9.6.3 AEC Tracking and DTD Sensitivity to Transmission Room Echo Path Changes

In this section, we show the influence that the nonlinear pre-processing of the transmission signal has on the convergence of the echo canceler to the true solution. This is done by shifting the position of the transmission room talker, thus changing the acoustic paths of the *transmission room*, $\mathbf{g}_p(n), p = 1, 2$ in Figure 9.1. Any increase of MSE because of this change is undesired and indicates that the channels have not been properly identified (see Chap. 5).

Figure 9.4 shows the MSE when we have a transmission room talker change at 5.2 seconds for unprocessed speech data ($\alpha = 0$) and for data processed with a nonlinearity with a distortion parameter of $\alpha = 0.5$ (see Chap. 5). The algorithm has to reconverge after the transmission room change when $\alpha = 0$, while it is not as badly affected for $\alpha = 0.5$. Furthermore, in

Table 9.1 The two-channel PC double-talk detector and echo canceler with modifications used in all simulations. The parameter ϱ bounds the maximum allowed coherence between the channels

Spectral estimation

$$S'_{p,q}(m) = \lambda S'_{p,q}(m-1) + (1-\lambda)\, D_p^H(m)D_q(m), \quad p,q = 1,2$$

$$\tilde{S}_{p,p}(m) = S'_{p,p}(m) + \text{diag}\{\delta_{p,0}\ldots\delta_{p,2L-1}\}, \quad p = 1,2$$

$$S_p(m) = \tilde{S}_{p,p}(m) \times$$
$$\left[I_{2L \times 2L} - \varrho^2 \left[\tilde{S}_{1,1}(m)\tilde{S}_{2,2}(m) \right]^{-1} S'_{2,1}(m)S'_{1,2}(m) \right], \quad p,q = 1,2$$

$$K_1(m) = S_1^{-1}(m) \left[D_1^*(m) - \varrho S'_{1,2}(m)\tilde{S}_{2,2}^{-1}(m)\, D_2^*(m) \right], \quad 0 \le \varrho \le 1$$

$$K_2(m) = S_2^{-1}(m) \left[D_2^*(m) - \varrho S'_{2,1}(m)\tilde{S}_{1,1}^{-1}(m)\, D_1^*(m) \right], \quad 0 \le \varrho \le 1$$

Double-talk detector (Background filter)

$$\underline{e}_b(m) = \underline{y}(m) - G^{01} \left[D_1(m)\underline{\hat{h}}_{b,1}(m-1) + D_2(m)\underline{\hat{h}}_{b,2}(m-1) \right]$$

$$\underline{\hat{h}}_{b,p}(m) = \underline{\hat{h}}_{b,p}(m-1) + (1-\lambda_b)K_p(m)\underline{e}_b(m), \quad p = 1,2$$

$$s(m) = \lambda_b s(m-1) + (1-\lambda_b)\, D^H(m)\underline{y}(m)$$

$$\eta^2(m) = \left[\underline{\hat{h}}_{b,1}^H(m)\ \underline{\hat{h}}_{b,2}^H(m) \right] s(m)$$

$$\sigma_{\underline{y}}^2(m) = \lambda_b \sigma_{\underline{y}}^2(m-1) + (1-\lambda_b)\underline{y}^H(m)\underline{y}(m)$$

$$\xi(m) = \eta(m)/\sigma_{\underline{y}}(m) < T, \Rightarrow \text{ double-talk}, \mu' = 0$$

$$\xi(m) = \eta(m)/\sigma_{\underline{y}}(m) \ge T, \Rightarrow \text{ no double-talk}, \mu' = \mu(1-\lambda)$$

Echo canceler (Foreground filter)

$$e(m) = y(m) - W^{01}F^{-1} \left[D_1(m)\underline{\hat{h}}_1(m-1) + D_2(m)\underline{\hat{h}}_2(m-1) \right]$$

$$\underline{\hat{h}}_p(m) = \underline{\hat{h}}_p(m-1) + \frac{2\mu' s(m)}{\psi'_{\min}(m)} K_p(m)F\psi\left[e(m)\right], \quad p = 1,2$$

$$s(m+1) = \lambda_s s(m) + (1-\lambda_s)\frac{s(m)}{L\beta}\sum_{n=mL}^{mL+L-1}\psi\left[\frac{|e(n)|}{s(m)}\right],$$

this simulation, it is shown that the DTD output (along bottom line of plot) does not false alarm because of the transmission path changes.

In the following simulations, we use $\alpha = 0.5$ so that we always have a solution close to the true receiving room echo paths when the canceler has converged in a mean-square error sense. In practice, a smaller value of α could be chosen.

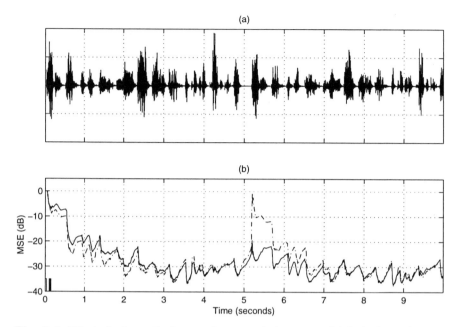

Fig. 9.4. Effect of echo path changes in transmission room. (a) Left channel transmission speech; the talker position changes at 5.2 seconds. (b) Mean-square error when echo paths of the transmission room changes: $\alpha = 0.5$ (Solid line), $\alpha = 0$ (Dash-dotted line)

9.6.4 AEC Tracking and DTD Sensitivity to Receiving Room Echo Path Changes

For the purpose of showing how the AEC/DTD system handles echo path changes in the *receiving room* we have chosen to study the reconvergence when the echo paths undergo one of four changes after the algorithm has been adapting for about 2 seconds. For simplicity, we assume that both echo paths undergo identical changes. Furthermore, we impose the same echo path change at two time instances so that variations of convergence rate due to input excitation variations are shown.

As mentioned in Chap. 6, there are infinite possibilities for echo path variations in reality. However, we have four cases that represent rather severe variations that may occur: time delay/advance which occur if a loudspeaker or microphone is repositioned, and echo path gain variation of +6 or −6 dB, which may occur if there is a corresponding increase or decrease in the loudspeaker volume. In order to have controlled echo path changes, the receiving room data have been manipulated such that we achieve the following changes at 2 and 5.2 seconds:

a)   Time delay: $h_p(n) \rightarrow h_p(n - 5)$, $p = 1, 2$.
b)   Time advance: $h_p(n) \rightarrow h_p(n + 14)$, $p = 1, 2$.

c)   6-dB increase of echo path gain: $h_p(n) \rightarrow 2h_p(n)$, $p = 1, 2$.

d)   6-dB decrease of echo path gain: $h_p(n) \rightarrow 0.5h_p(n)$, $p = 1, 2$.

Figure 9.5 shows the performance of the echo canceler for the echo path situations described above. At the first time the echo paths change (2 seconds), both algorithms reconverge at about the same rate. After the second echo path change at 5 seconds, the robust algorithm does not exploit the input speech as well as the non-robust algorithm and therefore shows slower reconvergence in this particular case.

Receiving room echo path changes do trigger false alarms in the double-talk detector (shown along bottom line of plots). These false alarms are due to the rather high threshold chosen[2], $T = 0.99$. By studying the statistic ξ (figures not shown here), it has been found that time delay/advance changes result in a *large decrease* of ξ over a *short period* of time while gain changes result in only a *slight decrease* of ξ but over a *longer period* of time. Choosing a lower threshold would decrease the sensitivity to echo path changes, particularly gain variations. In reality, a time shift of the whole impulse response is less probable. It is more likely that the direct path is unchanged while the tail changes slowly due to moving objects in the room. These changes will not result in as many false alarms as the case studied in this section.

9.6.5 AEC/DTD Performance During Double-Talk

The performance of the echo canceler during double-talk is presented in this section. In Fig. 9.6, we show the divergence that results from the presence of receiving room speech where we use either a male or female receiving room talker speaking the same phrase. For each talker, the speech sequence is repeated twice, so that two different instances of double-talk occur. The transmission room level was chosen such that EBR $= \sigma_{y_e}^2/(\sigma_v^2 + \sigma_w^2) \approx 1$ (0 dB).

As shown in Fig. 9.6, the PC DTD does not detect double-talk at the very beginning of the first burst at 4.2 seconds. This has the consequence that the non-robust algorithm starts to diverge, especially for the female talker, Fig. 9.6(d). The robust algorithm shows much slower divergence during this difficult situation. For the second double-talk burst, the PC DTD is accurate enough so that divergence is avoided completely for both robust and non-robust algorithms.

9.7 Conclusions

In this chapter, we have presented a double-talk detector based on a pseudo-coherence measure and a robust multichannel frequency-domain adaptive algorithm suitable for acoustic echo cancellation. Together, this DTD and the

---

[2] Note that in the figures, we choose to present the DTD results from the non-robust AEC/DTD configuration.

frequency-domain algorithm, make a system that shows good performance with respect to convergence rate and good behavior during double-talk, similar to what was found with the time-domain NCC DTD and FRLS canceler in Chap. 6. This system has been evaluated for the two-channel case using real-life recorded data. Significant results with the proposed system are:

- As with the NCC DTD, the PC DTD does not require any assumptions about the echo path, e.g., the echo path gain. This is an important property since the acoustic echo paths are highly variable.
- False alarms of the DTD, triggered by echo path variations, can be kept low by lowering the threshold. Possible divergence due to detection misses can be significantly reduced by using the robust version of the frequency-domain algorithm.
- Looking at the equations for the frequency-domain algorithm, Table 9.1, we find that they are based on computations involving simple arithmetic and Fourier transforms. Even though we have not investigated the dynamic range of the computations, these calculations can most likely, with some effort, be successfully implemented in a fixed-point signal processing platform and thus can provide a cost-efficient implementation of multichannel echo cancellation.

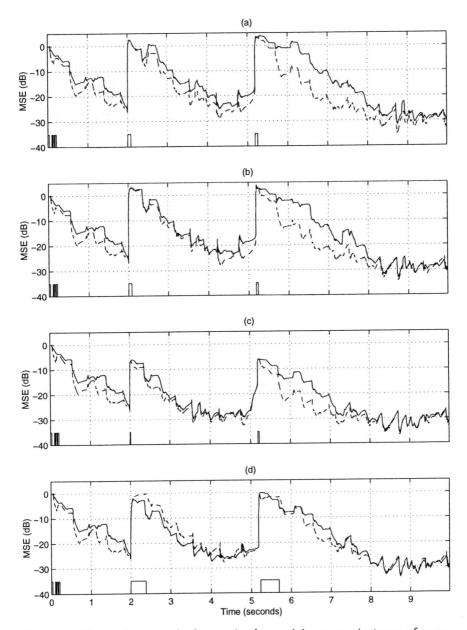

Fig. 9.5. Effects of echo path changes in the receiving room, in terms of mean-square error of the two-channel robust (solid) and non-robust (dashed) frequency-domain algorithms with the pseudo-coherence-based DTD. The echo paths in the receiving room change as: (a) Time delay of 5 samples. (b) Time advance of 14 samples. (c) 6-dB increase of echo path gain. (d) 6-dB decrease of echo path gain. The rectangles in the bottom of the figures indicate where double-talk has been detected (in this case, they are false alarms). These double-talk detection results were obtained for the threshold value $T = 0.99$ used with the non-robust algorithm

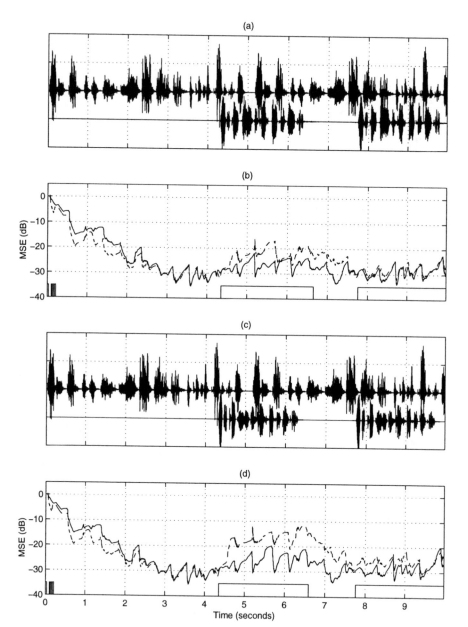

Fig. 9.6. Double-talk performance of the two-channel frequency-domain algorithm with pseudo-coherence-based DTD. (a) Left channel transmission room speech (upper), receiving room speech, male talker (lower). (b) Mean-square error of left channel, male talker. (c) Left channel transmission room speech (upper), receiving room speech, female talker (lower). (d) Mean-square error of left channel, female talker. Other conditions the same as in Fig. 9.5

10. Linear Interpolation Applied to Adaptive Filtering

10.1 Introduction

While linear prediction has been successfully applied to many topics in signal processing, linear interpolation has received little attention. This chapter gives some results on linear interpolation and shows that many well-known variables or equations can be formulated in terms of linear interpolation. Also, the so-called principle of orthogonality is generalized. From this theory, we then give a generalized least mean square algorithm and a generalized affine projection algorithm.

In the context of system identification, the error signal at time n between the system (which we assume to be linear and time-invariant) and model filter (finite impulse response) outputs is given by

$$e(n) = y(n) - \hat{y}(n) , \tag{10.1}$$

where

$$\hat{y}(n) = \hat{\mathbf{h}}^T \mathbf{x}(n)$$
$$= \sum_{k=0}^{K-1} \hat{\mathbf{h}}_k^T \mathbf{x}_k(n) \tag{10.2}$$

is an estimate of the output signal $y(n)$,

$$\hat{\mathbf{h}} = \left[\hat{h}_0 \; \hat{h}_1 \; \cdots \; \hat{h}_{L-1} \right]^T$$
$$= \left[\hat{\mathbf{h}}_0^T \; \hat{\mathbf{h}}_1^T \; \cdots \; \hat{\mathbf{h}}_{K-1}^T \right]^T ,$$

is the model filter, where

$$\hat{\mathbf{h}}_k = \left[\hat{h}_{kN} \; \hat{h}_{kN+1} \; \cdots \; \hat{h}_{kN+N-1} \right]^T , \quad k = 0, 1, ..., K-1 ,$$

and

$$\mathbf{x}(n) = \left[x(n) \; x(n-1) \; \cdots \; x(n-L+1) \right]^T$$
$$= \left[\mathbf{x}_0^T(n) \; \mathbf{x}_1^T(n) \; \cdots \; \mathbf{x}_{K-1}^T(n) \right]^T$$

is a vector containing $L = KN$ samples of the input signal x ending at the sample $x(n)$, where

$$\mathbf{x}_k(n) = \begin{bmatrix} x(n - kN) & x(n - kN - 1) & \cdots & x(n - kN - N + 1) \end{bmatrix}^T ,$$
$$k = 0, 1, ..., K - 1 ,$$

is an $N \times 1$ vector. Superscript ^{T} denotes transpose of a vector or a matrix. We have partitioned the filter $\hat{\mathbf{h}}$ of length L into K sub-filters $\hat{\mathbf{h}}_k$ of length N. For $K = 1$, there is only one sub-filter and it is the whole filter itself. For $N = 1$, we have as many sub-filters as the coefficients of the filter $\hat{\mathbf{h}}$. Partitioning will be used extensively in the rest of this chapter and, as we will see, it will help us to design new adaptive algorithms. Following this decomposition, all matrices and vectors will be partitioned in the same way.

The mean-square error criterion is defined as

$$J = E\left\{e^2(n)\right\} , \tag{10.3}$$

where $E\{\cdot\}$ is the statistical expectation operator. The minimization of (10.3) with respect to the filter leads to the Wiener-Hopf equation [66]:

$$\mathbf{R}\hat{\mathbf{h}} = \mathbf{r} , \tag{10.4}$$

where

$$\mathbf{R} = E\{\mathbf{x}(n)\mathbf{x}^T(n)\}$$
$$= \begin{bmatrix} \mathbf{R}_{0,0} & \mathbf{R}_{0,1} & \cdots & \mathbf{R}_{0,K-1} \\ \mathbf{R}_{1,0} & \mathbf{R}_{1,1} & \cdots & \mathbf{R}_{1,K-1} \\ \vdots & \vdots & \ddots & \vdots \\ \mathbf{R}_{K-1,0} & \mathbf{R}_{K-1,1} & \cdots & \mathbf{R}_{K-1,K-1} \end{bmatrix} \tag{10.5}$$

is the input signal covariance matrix, with $\mathbf{R}_{k,j} = E\{\mathbf{x}_k(n)\mathbf{x}_j^T(n)\}$, and

$$\mathbf{r} = E\{y(n)\mathbf{x}(n)\}$$
$$= \begin{bmatrix} \mathbf{r}_0^T & \mathbf{r}_1^T & \cdots & \mathbf{r}_{K-1}^T \end{bmatrix}^T \tag{10.6}$$

is the cross-correlation vector between the input and output signals, with $\mathbf{r}_k = E\{y(n)\mathbf{x}_k(n)\}$. Since $\mathbf{R}$ is symmetric, we have $\mathbf{R}_{k,j}^T = \mathbf{R}_{j,k}$.

Consider the interpolation error vectors:

$$\mathbf{z}_k(n) = \sum_{j=0}^{K-1} \mathbf{C}_{k,j}\mathbf{x}_j(n) \tag{10.7}$$

$$= \mathbf{x}_k(n) + \sum_{j=0,j\neq k}^{K-1} \mathbf{C}_{k,j}\mathbf{x}_j(n)$$

$$= \mathbf{x}_k(n) - \hat{\mathbf{x}}_k(n) ,$$
$$k = 0, 1, ..., K - 1 ,$$

with

$$\mathbf{C}_{k,k} = \mathbf{I}_{N \times N} \tag{10.8}$$

and

$$\hat{\mathbf{x}}_k(n) = - \sum_{j=0, j \neq k}^{K-1} \mathbf{C}_{k,j} \mathbf{x}_j(n) . \qquad (10.9)$$

Matrices $\mathbf{C}_{k,j}$ are the interpolators obtained by minimizing

$$J_k = E\left\{\mathbf{z}_k^T(n)\mathbf{z}_k(n)\right\} , \quad k = 0, 1, ..., K-1 . \qquad (10.10)$$

Matrix differentiation expresses the solution as a set of simultaneous equations

$$\sum_j \mathbf{C}_{k,j} \mathbf{R}_{j,l} = \mathbf{0}_{N \times N} , \quad l \neq k , \qquad (10.11)$$

which is also recognized as a generalized orthogonality principle.

A general factorization of $\mathbf{R}^{-1}$ can be stated as follows:

Lemma 1:

$$\mathbf{R}^{-1} = \begin{bmatrix} \mathbf{Q}_0^{-1} & \mathbf{0}_{N \times N} & \cdots & \mathbf{0}_{N \times N} \\ \mathbf{0}_{N \times N} & \mathbf{Q}_1^{-1} & \cdots & \mathbf{0}_{N \times N} \\ \vdots & \vdots & \ddots & \vdots \\ \mathbf{0}_{N \times N} & \mathbf{0}_{N \times N} & \cdots & \mathbf{Q}_{K-1}^{-1} \end{bmatrix} \begin{bmatrix} \mathbf{I}_{N \times N} & \mathbf{C}_{0,1} & \cdots & \mathbf{C}_{0,K-1} \\ \mathbf{C}_{1,0} & \mathbf{I}_{N \times N} & \cdots & \mathbf{C}_{1,K-1} \\ \vdots & \vdots & \ddots & \vdots \\ \mathbf{C}_{K-1,0} & \mathbf{C}_{K-1,1} & \cdots & \mathbf{I}_{N \times N} \end{bmatrix}$$

$$= \mathbf{Q}^{-1} \mathbf{C} , \qquad (10.12)$$

where

$$\mathbf{Q}_k = \sum_{j=0}^{K-1} \mathbf{C}_{k,j} \mathbf{R}_{j,k} , \quad k = 0, 1, ..., K-1 . \qquad (10.13)$$

Proof: The proof is rather straightforward by post-multiplying both sides of (10.12) by $\mathbf{R}$ and showing that the result on the right-hand side is equal to the identity matrix with the help of (10.11).

Since $\mathbf{R}^{-1}$ is symmetric, we have $\mathbf{Q}_k^{-1}\mathbf{C}_{k,j} = \mathbf{C}_{j,k}^T \mathbf{Q}_j^{-1}$. We have maximum factorization for $N = 1$ and no factorization for $K = 1$. For $N = 1$, this factorization is hinted at in [74] and we have:

$$\mathbf{R}^{-1} = \begin{bmatrix} 1/E\{z_0^2(n)\} & 0 & \cdots & 0 \\ 0 & 1/E\{z_1^2(n)\} & \cdots & 0 \\ \vdots & \vdots & \ddots & \vdots \\ 0 & 0 & \cdots & 1/E\{z_{K-1}^2(n)\} \end{bmatrix}$$

$$\times \begin{bmatrix} 1 & c_{0,1} & \cdots & c_{0,K-1} \\ c_{1,0} & 1 & \cdots & c_{1,K-1} \\ \vdots & \vdots & \ddots & \vdots \\ c_{K-1,0} & c_{K-1,1} & \cdots & 1 \end{bmatrix} , \qquad (10.14)$$

where the first and last lines of $\mathbf{R}^{-1}$ contain respectively the normalized forward and backward predictors and all the lines between contain the normalized interpolators. The main diagonal of $\mathbf{R}^{-1}$ contains the inverse of all the interpolation error energies, that can be computed recursively by using the Levinson algorithm [107]. Suppose K is even; since $\mathbf{R}^{-1}$ is persymmetric, we have $E\{z_k^2(n)\} = E\{z_{K-1-k}^2(n)\}$, $k = 0, 1, ..., K/2 - 1$ [74]. Obviously, the correlation of the input signal is strongly linked to the interpolation error energy: the higher the correlation, the smaller the interpolation error energy. In other words, the better the signal is interpolated, the higher the condition number of $\mathbf{R}$.

Definition: Let $\mathbf{A}$ be symmetric, positive definite, and let $\lambda_{\min}(\mathbf{A})$ and $\lambda_{\max}(\mathbf{A})$ denote the extreme eigenvalues of $\mathbf{A}$. Then by the *condition number* of $\mathbf{A}$ we mean

$$\text{cond}(\mathbf{A}) = \lambda_{\max}(\mathbf{A})/\lambda_{\min}(\mathbf{A}) . \tag{10.15}$$

It can be shown that:

$$\text{cond}(\mathbf{R}) \geq \text{cond}([\mathbf{Q}_k]_{(L/2)\times(L/2)}) \geq \cdots \geq \text{cond}([\mathbf{Q}_k]_{1\times1}) = 1 , \tag{10.16}$$

where $[\mathbf{Q}_k]_{M\times M}$ is the matrix $\mathbf{Q}_k$ of size $M \times M$.

We have some interesting orthogonality properties:

Lemma 2:

$$E\{\mathbf{z}_k(n)\mathbf{x}_j^T(n)\} = \mathbf{0}_{N\times N} , \ \forall k, j = 0, 1, ..., K - 1 , \ k \neq j , \tag{10.17}$$

$$\text{and hence} : E\{\mathbf{x}_k^T(n)\mathbf{z}_j(n)\} = 0 . \tag{10.18}$$

Proof: The proof is straightforward from (10.11).

We deduce from Lemma 2 the following properties:

$$E\{\mathbf{z}_k(n)\mathbf{z}_k^T(n)\} = E\{\mathbf{z}_k(n)\mathbf{x}_k^T(n)\} = \mathbf{Q}_k , \tag{10.19}$$

$$E\{\mathbf{z}_k(n)\mathbf{z}_j^T(n)\} = E\{\mathbf{z}_k(n)\mathbf{x}_k^T(n)\}\mathbf{C}_{j,k}^T = \mathbf{Q}_k\mathbf{C}_{j,k}^T , \tag{10.20}$$

we have a factorization similar to (10.12),

$$\begin{aligned}
\mathbf{R}_z &= E\{\mathbf{z}(n)\mathbf{z}^T(n)\} \\
&= \begin{bmatrix} \mathbf{Q}_0 & \mathbf{0}_{N\times N} & \cdots & \mathbf{0}_{N\times N} \\ \mathbf{0}_{N\times N} & \mathbf{Q}_1 & \cdots & \mathbf{0}_{N\times N} \\ \vdots & \vdots & \ddots & \vdots \\ \mathbf{0}_{N\times N} & \mathbf{0}_{N\times N} & \cdots & \mathbf{Q}_{K-1} \end{bmatrix} \begin{bmatrix} \mathbf{I}_{N\times N} & \cdots & \mathbf{C}_{K-1,0}^T \\ \mathbf{C}_{0,1}^T & \cdots & \mathbf{C}_{K-1,1}^T \\ \vdots & \ddots & \vdots \\ \mathbf{C}_{0,K-1}^T & \cdots & \mathbf{I}_{N\times N} \end{bmatrix} \\
&= \mathbf{Q}\mathbf{C}^T , \tag{10.21}
\end{aligned}$$

where

$$\begin{aligned}
\mathbf{z}(n) &= \begin{bmatrix} \mathbf{z}_0^T(n) & \mathbf{z}_1^T(n) & \cdots & \mathbf{z}_{K-1}^T(n) \end{bmatrix}^T \\
&= \mathbf{C}\mathbf{x}(n) ,
\end{aligned}$$

and finally:

$$\mathbf{R}^{-1} = \mathbf{Q}^{-1}\mathbf{R}_z\mathbf{Q}^{-1} . \tag{10.22}$$

Since $\mathbf{Q}^{-1}\mathbf{R}^{-1}\mathbf{Q}$ and $\mathbf{R}^{-1}$ have the same spectrum, $\mathbf{R}^{-1}$ and $\mathbf{Q}^{-2}\mathbf{R}_z$ (or $\mathbf{R}_z\mathbf{Q}^{-2}$) also have the same spectrum.

Let $\lambda_k(\mathbf{R})$ denote the kth eigenvalue of $\mathbf{R}$, where we order the eigenvalues in increasing order. Since $\mathbf{Q}^{-2}$ and $\mathbf{R}_z$ are symmetric and positive definite matrices, we have for $N = 1$,

$$\min_j \frac{1}{E^2\{z_j^2(n)\}} \le \frac{\lambda_k(\mathbf{R}^{-1})}{\lambda_k(\mathbf{R}_z)} \le \max_j \frac{1}{E^2\{z_j^2(n)\}} . \tag{10.23}$$

Equation (10.22) shows another way to factor the inverse of the input signal covariance matrix. We see that this inverse depends explicitly on the interpolation error covariance matrix and the inverse energies of the interpolators, so that each element of the matrix $\mathbf{R}^{-1}$ can be written (for $N = 1$) as:

$$[\mathbf{R}^{-1}]_{k,j} = \frac{E\{z_k(n)z_j(n)\}}{E\{z_k^2(n)\}E\{z_j^2(n)\}} . \tag{10.24}$$

It is interesting to compare (10.24) to the normalized cross-correlation coefficient of the input signals:

$$r_{x_k x_j} = \frac{E\{x_k(n)x_j(n)\}}{\sqrt{E\{x_k^2(n)\}E\{x_j^2(n)\}}} . \tag{10.25}$$

Example: $K = 2$: In this case, we have:

$$\mathbf{z}_0(n) = \mathbf{x}_0(n) + \mathbf{C}_{0,1}\mathbf{x}_1(n) , \tag{10.26}$$

$$\mathbf{z}_1(n) = \mathbf{x}_1(n) + \mathbf{C}_{1,0}\mathbf{x}_0(n) , \tag{10.27}$$

where from (10.11)

$$\mathbf{C}_{0,1} = -\mathbf{R}_{0,1}\mathbf{R}_{1,1}^{-1} , \tag{10.28}$$

$$\mathbf{C}_{1,0} = -\mathbf{R}_{1,0}\mathbf{R}_{0,0}^{-1} , \tag{10.29}$$

are the two interpolators obtained by minimizing $E\{\mathbf{z}_0^T(n)\mathbf{z}_0(n)\}$ and $E\{\mathbf{z}_1^T(n)\mathbf{z}_1(n)\}$. Hence:

$$\mathbf{R}^{-1} = \begin{bmatrix} \mathbf{Q}_0^{-1} & \mathbf{0}_{N\times N} \\ \mathbf{0}_{N\times N} & \mathbf{Q}_1^{-1} \end{bmatrix} \begin{bmatrix} \mathbf{I}_{N\times N} & -\mathbf{R}_{0,1}\mathbf{R}_{1,1}^{-1} \\ -\mathbf{R}_{1,0}\mathbf{R}_{0,0}^{-1} & \mathbf{I}_{N\times N} \end{bmatrix} , \tag{10.30}$$

where

$$\mathbf{Q}_0 = \mathbf{R}_{0,0} - \mathbf{R}_{0,1}\mathbf{R}_{1,1}^{-1}\mathbf{R}_{1,0} , \tag{10.31}$$

$$\mathbf{Q}_1 = \mathbf{R}_{1,1} - \mathbf{R}_{1,0}\mathbf{R}_{0,0}^{-1}\mathbf{R}_{0,1} , \tag{10.32}$$

are the interpolation error energy covariance matrices or the *Schur complements* of $\mathbf{R}$ with respect to $\mathbf{R}_{1,1}$ and $\mathbf{R}_{0,0}$ [105].

Interpretation of the Kalman Gain:

Let us define the Kalman gain,

$$\mathbf{g}(n) = \mathbf{R}^{-1}\mathbf{x}(n) \tag{10.33}$$

$$= \begin{bmatrix} \mathbf{g}_0^T(n) & \mathbf{g}_1^T(n) & \cdots & \mathbf{g}_{K-1}^T(n) \end{bmatrix}^T$$

$$= \begin{bmatrix} g_0(n) & g_1(n) & \cdots & g_{L-1}(n) \end{bmatrix}^T ,$$

and the maximum likelihood variable,

$$\theta(n) = \mathbf{x}^T(n)\mathbf{R}^{-1}\mathbf{x}(n) \,, \tag{10.34}$$

we have:

$$\mathbf{g}_k(n) = \mathbf{Q}_k^{-1}\mathbf{z}_k(n) \,, \quad \forall k = 0, 1, ..., K-1 \,, \tag{10.35}$$

$$\theta(n) = \sum_{k=0}^{K-1} \mathbf{x}_k^T(n)\mathbf{Q}_k^{-1}\mathbf{z}_k(n) \,. \tag{10.36}$$

For maximum factorization ($N = 1$), each element $g_k(n)$ of the Kalman gain is the interpolation error of $x(n-k)$ from its past and its future normalized by its corresponding energy, i.e. $g_k(n) = z_k(n)/E\{z_k^2(n)\}$.

Principle of Orthogonality:

One of the most important theorems in Wiener theory is the principle of orthogonality [66]

$$E\{\mathbf{x}(n)e(n)\} = \mathbf{0}_{L\times 1} \,, \tag{10.37}$$

which states that the necessary and sufficient condition for the cost function $J = E\{e^2(n)\}$ to attain its minimum value is that the corresponding value of the estimation error $e(n)$ (with the optimal filter) is orthogonal to each input sample $x(n-k)$. We propose to generalize this to:

Lemma 3:

$$E\{\mathbf{z}(n)e(n)\} = \mathbf{0}_{L\times 1} \,. \tag{10.38}$$

Proof: Obvious from (10.1)-(10.7).

Thus, by using (10.38), we redefine the Wiener-Hopf equation as:

$$\mathbf{Q}\hat{\mathbf{h}} = \mathbf{q} \,, \tag{10.39}$$

where $\mathbf{q} = E\{y(n)\mathbf{z}(n)\}$ and $\mathbf{Q}$ is a block diagonal matrix in the general case or a diagonal matrix for the maximum factorization case ($N = 1$). The optimal solution (for $N = 1$) can be seen as the normalized cross-correlation vector between the system output y and the interpolation error vector $\mathbf{z}$.

10.2 A Generalized Least Mean Square (GLMS) Algorithm

According to the steepest descent method, the updated value of $\hat{\mathbf{h}}$ at time n is computed by using the principle of orthogonality with the simple recursive relation [66]:

$$\hat{\mathbf{h}}(n) = \hat{\mathbf{h}}(n-1) + \mu_x E\{\mathbf{x}(n)e(n)\} \,, \tag{10.40}$$

with

$$e(n) = y(n) - \hat{\mathbf{h}}^T(n-1)\mathbf{x}(n) \; ; \tag{10.41}$$

and the classical stochastic approximation (consisting of approximating the gradient with its instantaneous value) [134] provides the LMS algorithm,

$$\hat{\mathbf{h}}(n) = \hat{\mathbf{h}}(n-1) + \mu_x \mathbf{x}(n) e(n) \; , \tag{10.42}$$

of which the classical mean weight convergence condition under the appropriate independence assumption is:

$$0 < \mu_x < \frac{2}{N\sigma_x^2} \; , \tag{10.43}$$

where the σ_x^2 is the power of the input signal. When this condition is satisfied, the weight vector converges in mean to the optimal Wiener-Hopf solution. However, it is well-known that the convergence of this algorithm depends on the condition number of the auto-correlation matrix $\mathbf{R}$ and may converge very slowly to the optimal solution with input signals like speech.

We propose here, by using Lemma 3, a generalized steepest descent algorithm:

$$\hat{\mathbf{h}}(n) = \hat{\mathbf{h}}(n-1) + \mu_z E\{\mathbf{z}(n) e(n)\} \; , \tag{10.44}$$

and a generalized LMS algorithm:

$$\hat{\mathbf{h}}(n) = \hat{\mathbf{h}}(n-1) + \mu_z \mathbf{z}(n) e(n) \; , \tag{10.45}$$

with

$$0 < \mu_z < \frac{2}{N\sigma_z^2} \tag{10.46}$$

and where σ_z^2 is the power of the signal z.

Define the weight-error vector

$$\boldsymbol{\varepsilon}(n) = \mathbf{h} - \hat{\mathbf{h}}(n) \; , \tag{10.47}$$

where $\mathbf{h}$ is the linear filter to be identified. Using the independence assumption, we easily see from (10.45) that

$$E\{\boldsymbol{\varepsilon}_k(n)\} = [\mathbf{I}_{N \times N} - \mu_z \mathbf{Q}_k] E\{\boldsymbol{\varepsilon}_k(n-1)\} \; , \quad k = 0, 1, ..., K-1 \; . \tag{10.48}$$

Equation (10.48) implies that the convergence of the algorithm depends on the condition number of the matrices $\mathbf{Q}_k$. For maximum factorization ($N = 1$), the condition number is equal to 1 and this algorithm is simply the classical Newton algorithm. For no factorization ($K = 1$), we recognize, of course, the classical LMS. For $N \neq 1$ and $K \neq 1$, we have a new class of adaptive algorithms with a convergence rate that, because of (10.16), lies between LMS and Newton algorithms.

Example 1: $K = 2$:

We deduce the following normalized algorithm:

$$\hat{\mathbf{h}}_k(n) = \hat{\mathbf{h}}_k(n-1) + \mu_z(n)\mathbf{z}_k(n)e(n) \ , \ \ k = 0, 1 \ , \tag{10.49}$$

where

$$\mu_z(n) = \frac{1}{\mathbf{z}_0^T(n)\mathbf{z}_0(n) + \mathbf{z}_1^T(n)\mathbf{z}_1(n)} \ , \tag{10.50}$$

and

$$\mathbf{z}_0(n) = \mathbf{x}(n) - \mathbf{R}_{0,1}\mathbf{R}_{1,1}^{-1}\mathbf{x}(n-N) \ , \tag{10.51}$$

$$\mathbf{z}_1(n) = \mathbf{x}(n-N) - \mathbf{R}_{1,0}\mathbf{R}_{0,0}^{-1}\mathbf{x}(n) \ . \tag{10.52}$$

Example 2: $N = 1$:

We have the Newton algorithm:

$$\hat{h}_k(n) = \hat{h}_k(n-1) + \frac{z_k(n)}{E\{z_k^2(n)\}}e(n) \ , \ \ k = 0, 1, ..., K-1 \ . \tag{10.53}$$

10.3 A Generalized Affine Projection Algorithm (GAPA)

Define a set of I *a priori* errors and I *a posteriori* errors:

$$\mathbf{e}(n) = \mathbf{y}(n) - \mathbf{X}^T(n)\hat{\mathbf{h}}(n-1) \ , \tag{10.54}$$

$$\mathbf{e}_a(n) = \mathbf{y}(n) - \mathbf{X}^T(n)\hat{\mathbf{h}}(n) \ , \tag{10.55}$$

where

$$\mathbf{y}(n) = \begin{bmatrix} y(n) \ y(n-1) \ \cdots \ y(n-I+1) \end{bmatrix}^T \ ,$$

$$\mathbf{e}(n) = \begin{bmatrix} e(n) \ e(n-1) \ \cdots \ e(n-I+1) \end{bmatrix}^T \ ,$$

$$\mathbf{e}_a(n) = \begin{bmatrix} e_a(n) \ e_a(n-1) \ \cdots \ e_a(n-I+1) \end{bmatrix}^T \ ,$$

and

$$\mathbf{X}(n) = \begin{bmatrix} \mathbf{X}_{I,0}^T(n) \ \mathbf{X}_{I,1}^T(n) \ \cdots \ \mathbf{X}_{I,K-1}^T(n) \end{bmatrix}^T$$

is a matrix of size $KN \times I$ with

$$\mathbf{X}_{I,k}(n) = \begin{bmatrix} \mathbf{x}_k(n) \ \mathbf{x}_k(n-1) \ \cdots \ \mathbf{x}_k(n-I+1) \end{bmatrix} \ , \ \ k = 0, 1..., K-1 \ ,$$

of size $N \times I$.

Using (10.54) and (10.55) plus the requirement that $\mathbf{e}_a(n) = \mathbf{0}_{I \times 1}$, we obtain:

$$\mathbf{X}^T(n)\Delta\hat{\mathbf{h}}(n) = \mathbf{e}(n) \ , \tag{10.56}$$

where

$$\Delta\hat{\mathbf{h}}(n) = \hat{\mathbf{h}}(n) - \hat{\mathbf{h}}(n-1)$$
$$= \left[\Delta\hat{\mathbf{h}}_0^T(n)\ \Delta\hat{\mathbf{h}}_1^T(n)\ \cdots\ \Delta\hat{\mathbf{h}}_{K-1}^T(n)\right]^T .$$

Equation (10.56) (I equations in KN unknowns, $I \leq KN$) is an underdetermined set of linear equations. Hence, it has an infinite number of solutions, out of which the minimum-norm solution is chosen so that the adaptive filter has smooth variations. This results in

$$\hat{\mathbf{h}}(n) = \hat{\mathbf{h}}(n-1) + \mathbf{X}(n)\left[\mathbf{X}^T(n)\mathbf{X}(n)\right]^{-1}\mathbf{e}(n) , \tag{10.57}$$

which is the classical APA [106].

Example: $K = 2$:

Consider two constraints that will be justified later on:

$$\mathbf{X}_{M,1}^T(n)\Delta\hat{\mathbf{h}}_0(n) = \mathbf{0}_{M\times 1} , \tag{10.58}$$

$$\mathbf{X}_{M,0}^T(n)\Delta\hat{\mathbf{h}}_1(n) = \mathbf{0}_{M\times 1} . \tag{10.59}$$

The new set of linear equations characterizing the generalized APA is:

$$\begin{bmatrix} \mathbf{X}_{I,0}^T(n) & \mathbf{X}_{I,1}^T(n) \\ \mathbf{X}_{M,1}^T(n) & \mathbf{0}_{M\times N} \\ \mathbf{0}_{M\times N} & \mathbf{X}_{M,0}^T(n) \end{bmatrix} \begin{bmatrix} \Delta\hat{\mathbf{h}}_0(n) \\ \Delta\hat{\mathbf{h}}_1(n) \end{bmatrix} = \begin{bmatrix} \mathbf{e}(n) \\ \mathbf{0}_{M\times 1} \\ \mathbf{0}_{M\times 1} \end{bmatrix} , \tag{10.60}$$

where M, $1 \leq M \leq I$, is an arbitrary chosen projection order and $\mathbf{X}_{M,k}(n)$ is defined similarly to $\mathbf{X}_{I,k}(n)$. If we suppose that $I+2M \leq 2N$ (more unknowns than equations), taking the minimum-norm solution of (10.60) yields:

$$\Delta\hat{\mathbf{h}}_0(n) = \mathbf{Z}_0(n)\left[\mathbf{Z}_0^T(n)\mathbf{Z}_0(n) + \mathbf{Z}_1^T(n)\mathbf{Z}_1(n)\right]^{-1}\mathbf{e}(n) , \tag{10.61}$$

$$\Delta\hat{\mathbf{h}}_1(n) = \mathbf{Z}_1(n)\left[\mathbf{Z}_0^T(n)\mathbf{Z}_0(n) + \mathbf{Z}_1^T(n)\mathbf{Z}_1(n)\right]^{-1}\mathbf{e}(n) , \tag{10.62}$$

where $\mathbf{Z}_k(n)$ (of size $N \times I$) is the projection of $\mathbf{X}_{I,k}(n)$ onto a subspace orthogonal to $\mathbf{X}_{M,j}(n)$, $k \neq j$, i.e.,

$$\mathbf{Z}_k(n) = \left\{\mathbf{I}_{N\times N} - \mathbf{X}_{M,j}(n)\left[\mathbf{X}_{M,j}^T(n)\mathbf{X}_{M,j}(n)\right]^{-1}\mathbf{X}_{M,j}^T(n)\right\}\mathbf{X}_{I,k}(n) ,$$
$$k,j = 0,1 , \ k \neq j . \tag{10.63}$$

We deduce the following orthogonality conditions:

$$\mathbf{X}_{M,j}^T(n)\mathbf{Z}_k(n) = \mathbf{0}_{M\times I} , \ k,j = 0,1 , \ k \neq j , \tag{10.64}$$

similar to Lemma 2, so that the above constraints have a sense.

General Case:

First, define the following matrix of size $N \times (K-1)M$:

$$\mathbf{X}_{g,k}(n) = \left[\mathbf{X}_{M,0}(n)\ \cdots\ \mathbf{X}_{M,k-1}(n)\ \mathbf{X}_{M,k+1}(n)\ \cdots\ \mathbf{X}_{M,K-1}(n)\right] ,$$
$$k = 0,1,...,K-1 .$$

The K orthogonality constraints are:

$$\mathbf{X}_{g,k}^T(n)\Delta\hat{\mathbf{h}}_k(n) = \mathbf{0}_{(K-1)M\times 1} , \quad k = 0, 1, ..., K - 1 , \qquad (10.65)$$

and by using the same steps as for $K = 2$, a solution similar to (10.61), (10.62) is obtained [supposing that $I + K(K - 1)M \leq KN$]:

$$\Delta\hat{\mathbf{h}}_k(n) = \mathbf{Z}_k(n) \left[\sum_{j=0}^{K-1} \mathbf{Z}_j^T(n)\mathbf{Z}_j(n) \right]^{-1} \mathbf{e}(n) , \qquad (10.66)$$
$$k = 0, 1, ..., K - 1 ,$$

where $\mathbf{Z}_k(n)$ is the projection of $\mathbf{X}_{M,k}(n)$ onto a subspace orthogonal to $\mathbf{X}_{g,k}(n)$, i.e.,

$$\mathbf{Z}_k(n) = \left\{ \mathbf{I}_{N\times N} - \mathbf{X}_{g,k}(n) \left[\mathbf{X}_{g,k}^T(n)\mathbf{X}_{g,k}(n) \right]^{-1} \mathbf{X}_{g,k}^T(n) \right\} \mathbf{X}_{I,k}(n) ,$$
$$k = 0, 1, ..., K - 1 . \qquad (10.67)$$

Note that this equation holds only under the condition $N \geq (K - 1)M$, so that the matrix that appears in (10.67) is invertible.

We can easily see that:

$$\mathbf{X}_{g,k}^T(n)\mathbf{Z}_k(n) = \mathbf{0}_{(K-1)M\times I} , \quad k = 0, 1, ..., K - 1 , \qquad (10.68)$$

which are the orthogonality properties in the context of projections.

10.4 Conclusions

The main contribution of this chapter was to show the importance of linear interpolation with regards to some interpretations and derivations of new adaptive algorithms. We have shown that many well-known variables, equations, and adaptive algorithms can be understood from a linear interpolation point of view. For example, the update of the Newton algorithm is simply the normalized interpolation error signal multiplied by the error signal between the system and the model. Thanks to this new formulation, we were able to find a new class of adaptive algorithms for both stochastic and projection algorithms. We believe that much more can be done and some other interpretations can be deduced, especially in the frequency domain.

References

1. M. Ali, "Stereophonic echo cancellation system using time-varying all-pass filtering for signal decorrelation," in *Proc. IEEE ICASSP*, 1998, pp. 3689–3692.
2. C. Antweiler, J. Grunwald, and H. Quack, "Approximation of optimal step-size control for acoustic echo cancellation," in *Proc. IEEE ICASSP*, 1997, pp. 295–298.
3. S. P. Applebaum, "Adaptive arrays," *Syracuse Univ. Res. Corp.*, Rep. SPL-709, June 1964; reprinted in *IEEE Trans. Antennas Propagat.*, vol. 24, pp. 573–598, Sept. 1976.
4. M. G. Bellanger, *Adaptive Digital Filters and Signal Analysis*. Marcel Dekker, New York, 1987.
5. J. Benesty, F. Amand, A. Gilloire, and Y. Grenier, "Adaptive filtering algorithms for stereophonic acoustic echo cancellation," in *Proc. IEEE ICASSP*, 1995, pp. 3099–3102.
6. J. Benesty and P. Duhamel, "Fast constant modulus adaptive algorithm," *IEE Proc.*, Pt. F, vol. 138, pp. 379–387, Aug. 1991.
7. J. Benesty and P. Duhamel, "A fast exact least mean square adaptive algorithm," *IEEE Trans. Signal Processing*, vol. 40, pp. 2904–2920, Dec. 1992.
8. J. Benesty, P. Duhamel, and Y. Grenier, "Multi-channel adaptive filtering applied to multi-channel acoustic echo cancellation," in *Proc. EUSIPCO*, 1996, pp. 1405–1408.
9. J. Benesty, P. Duhamel, and Y. Grenier, "A multi-channel affine projection algorithm with applications to multi-channel acoustic echo cancellation," *IEEE Signal Processing Lett.*, vol. 3, pp. 35–37, Feb. 1996.
10. J. Benesty, T. Gänsler, and P. Eneroth, "Multi-channel sound, acoustic echo cancellation, and multi-channel time-domain adaptive filtering," in *Acoustic Signal Processing for Telecommunication*, S. L. Gay and J. Benesty, eds., Kluwer Academic Publishers, 2000, chap. 6, pp. 101–120.
11. J. Benesty, A. Gilloire, and Y. Grenier, "A frequency domain stereophonic acoustic echo canceler exploiting the coherence between the channels," *J. Acoust. Soc. Am.*, vol. 106, pp. L30–L35, Sept. 1999.
12. J. Benesty, A. Gilloire, and Y. Grenier, "A frequency-domain stereophonic acoustic echo canceler exploiting the coherence between the channels and using nonlinear transformations," in *Proc. IWAENC*, 1999, pp. 28–31.
13. J. Benesty, S. W. Li, and P. Duhamel, "A gradient-based adaptive algorithm with reduced complexity, fast convergence and good tracking characteristics," in *Proc. IEEE ICASSP*, 1992, pp. 5–8.
14. J. Benesty and D. R. Morgan, "Frequency-domain adaptive filtering revisited, generalization to the multi-channel case, and application to acoustic echo cancellation," in *Proc. IEEE ICASSP*, 2000, pp. 789–792.

15. J. Benesty and D. R. Morgan, "Multi-channel frequency-domain adaptive filtering," in *Acoustic Signal Processing for Telecommunication*, S. L. Gay and J. Benesty, eds., Kluwer Academic Publishers, 2000, chap. 7, pp. 121–133.

16. J. Benesty, D. R. Morgan, and J. H. Cho, "A new class of doubletalk detectors based on cross-correlation," *IEEE Trans. Speech Audio Processing*, vol. 8, pp. 168–172, Mar. 2000.

17. J. Benesty, D. R. Morgan, J. L. Hall, and M. M. Sondhi, "Stereophonic acoustic echo cancellation using nonlinear transformations and comb filtering," in *Proc. IEEE ICASSP*, 1998, pp. 3673–3676.

18. J. Benesty, D. R. Morgan, J. L. Hall, and M. M. Sondhi, "Synthesized stereo combined with acoustic echo cancellation for desktop conferencing," *Bell Labs Tech. J.*, vol. 3, pp. 148–158, July-Sept. 1998.

19. J. Benesty, D. R. Morgan, and M. M. Sondhi, "A better understanding and an improved solution to the specific problems of stereophonic acoustic echo cancellation," *IEEE Trans. Speech Audio Processing*, vol. 6, pp. 156–165, Mar. 1998.

20. J. Benesty, D. R. Morgan, and M. M. Sondhi, "A hybrid mono/stereo acoustic echo canceler," *IEEE Trans. Speech Audio Processing*, vol. 6, pp. 468–475, Sept. 1998.

21. D. A. Berkley and J. L. Flanagan, "HuMaNet: an experimental human-machine communications network based on ISDN wideband audio," *AT&T Tech. J.*, vol. 69, pp. 87–99, Sept./Oct. 1990.

22. J. Blauert, *Spatial Hearing*. MIT Press, Cambridge, MA, 1983.

23. D. H. Brandwood, "A complex gradient operator and its application in adaptive array theory," *IEE Proc.*, Pts. F and H, vol. 130, pp. 11–16, Feb. 1983.

24. C. Breining, P. Dreiseitel, E. Hänsler, A. Mader, B. Nitsch, H. Puder, T. Schertler, G. Schmidt, and J. Tilp, "Acoustic echo control—An application of very-high-order adaptive filters," *IEEE Signal Processing Mag.*, vol. 16, pp. 42–69, July 1999.

25. K. Bullington and J. M. Fraser, "Engineering aspects of TASI," *Bell Syst. Tech. J.*, pp. 353–364, Mar. 1959.

26. J. Chen, "3D audio and virtual acoustical environment synthesis," in *Acoustic Signal Processing for Telecommunication*, S. L. Gay and J. Benesty, eds., Kluwer Academic Publishers, 2000, chap. 13, pp. 283–301.

27. J. Chen, H. Bes, J. Vandewalle, and P. Janssens, "A new structure for sub-band acoustic echo canceler," in *Proc. IEEE ICASSP*, 1988, pp. 2574–2577.

28. J. H. Cho, D. R. Morgan, and J. Benesty, "An objective technique for evaluating doubletalk detectors in acoustic echo cancelers," *IEEE Trans. Speech Audio Processing*, vol. 7, pp. 718–724, Nov. 1999.

29. J. Cioffi and T. Kailath, "Fast, recursive-least-squares transversal filters for adaptive filtering," *IEEE Trans. Acoust., Speech, Signal Processing*, vol. 34, pp. 304–337, Apr. 1984.

30. P. M. Clarkson, *Optimal and Adaptive Signal Processing*. CRC, London, 1993.

31. R. E. Crochiere and L. R. Rabiner, *Multirate digital signal processing*. Prentice-Hall, Englewood Cliffs, NJ, 1983.

32. P. L. DeLeon and D. M. Etter, "Experimental results with increased bandwidth analysis filters in oversampled, subband acoustic echo cancelers," *IEEE Signal Processing Lett.*, vol. 2, pp. 1–3, Jan. 1995.

33. M. Dentino, J. McCool, and B. Widrow, "Adaptive filtering in the frequency domain," *Proc. IEEE*, vol. 66, pp. 1658–1659, Dec. 1978.

34. E. J. Diethorn, "Perceptually optimum adaptive filter tap profiles for subband acoustic echo cancellers," in *Proc. IEEE ASSP Workshop on Applications of Signal Processing to Audio and Acoustics*, 1995.

35. E. J. Diethorn, "An algorithm for subband echo suppression in speech communications," *Private Communication*, 1998.

36. N. I. Durlach and H. S. Colburn, "Binaural phenomena," in *Handbook of Perception, Volume IV, Hearing*, E. C. Carterette and M. P. Friedman, eds., Academic Press, New York, 1978, chap. 10.

37. D. L. Duttweiler, "A twelve-channel digital echo canceler," *IEEE Trans. Commun.*, vol. 26, pp. 647–653, May 1978.

38. D. L. Duttweiler, "Proportionate normalized least-mean-squares adaptation in echo cancelers," *IEEE Trans. Speech Audio Processing*, vol. 8, pp. 508–518, Sept. 2000.

39. D. L. Duttweiler and Y. S. Chen, "A single chip VLSI echo canceller," *Bell Syst. Tech. J.*, vol. 59, pp. 149–160, Feb. 1980.

40. P. Eneroth, J. Benesty, T. Gänsler, and S. L. Gay, "Comparison of different adaptive algorithms for stereophonic acoustic echo cancellation," in *Proc. EUSIPCO*, 2000, pp. 1835–1837.

41. P. Eneroth, S. L. Gay, T. Gänsler, and J. Benesty, "A real-time stereophonic acoustic subband echo canceler," in *Acoustic Signal Processing for Telecommunication*, S. L. Gay and J. Benesty, eds., Kluwer Academic Publishers, 2000, chap. 8, pp. 135–152.

42. E. Eweda, "Comparison of RLS, LMS, and sign algorithms for tracking randomly time-varying channels," *IEEE Trans. Signal Processing*, vol. 42, pp. 2937–2944, Nov. 1994.

43. D. D. Falconer and L. Ljung, "Application of fast Kalman estimation to adaptive equalization," *IEEE Trans. Commun.*, vol. 26, pp. 1439–1446, Oct. 1978.

44. B. Farhang-Boroujeny and Z. Wang, "Adaptive filtering in subbands: Design issues and experimental results for acoustic echo cancellation," *Signal Processing*, vol. 61, pp. 213–223, 1997.

45. E. R. Ferrara, Jr., "Fast implementation of LMS adaptive filter," *IEEE Trans. Acoust., Speech, Signal Processing*, vol. 28, pp. 474–475, Aug. 1980.

46. A. Feuer and E. Weinstein, "Convergence analysis of LMS filters with uncorrelated gaussian data," *IEEE Trans. Acoust., Speech, Signal Processing*, vol. 33, pp. 222–230, Feb. 1985.

47. I. Furukawa, "A design of canceller of broad band acoustic echo," in *Int. Teleconference Symposium*, 1984, pp. 1/8–8/8.

48. T. Gänsler, "A double-talk resistant subband echo canceller," *Signal Processing*, vol. 65, pp. 89–101, Feb. 1998.

49. T. Gänsler and J. Benesty, "Stereophonic acoustic echo cancellation and two-channel adaptive filtering: an overview," *Int. J. Adapt. Control Signal Process.*, vol. 14, pp. 565–586, Sept. 2000.

50. T. Gänsler and P. Eneroth, "Influence of audio coding on stereophonic acoustic echo cancellation," in *Proc. IEEE ICASSP*, 1998, pp. 3649–3652.

51. T. Gänsler, S. L. Gay, M. M. Sondhi, and J. Benesty, "Double-talk robust fast converging algorithms for network echo cancellation," *IEEE Trans. Speech Audio Processing*, vol. 8, pp. 656–663, Nov. 2000.

52. T. Gänsler, M. Hansson, C.-J. Ivarsson, and G. Salomonsson, "A double-talk detector based on coherence," *IEEE Trans. Commun.*, vol. 44, pp. 1421–1427, Nov. 1996.

53. M. B. Gardner, "Historical background of the Haas and/or precedence effect," *Acoust. Soc. Am.*, vol. 43, pp. 1243–1248, 1968.

54. S. L. Gay, "An efficient, fast converging adaptive filter for network echo cancellation," in *Proc. of Asilomar*, Nov. 1998.

55. S. L. Gay, "The fast affine projection algorithm," in *Acoustic Signal Processing for Telecommunication*, S. L. Gay and J. Benesty, eds., Kluwer Academic Publishers, 2000, chap. 2, pp. 23–45.

56. S. L. Gay and S. Tavathia, "The fast affine projection algorithm," in *Proc. IEEE ICASSP*, 1995, pp. 3023–3026.

57. O. Ghitza, "Auditory models and human performance in tasks related to speech coding and speech recognition," *IEEE Trans. Speech Audio Processing*, vol. 2, pp. 115–132, Jan. 1994.

58. A. Gilloire and V. Turbin, "Using auditory properties to improve the behavior of stereophonic acoustic echo cancellers," in *Proc. IEEE ICASSP*, 1998, pp. 3681–3684.

59. A. Gilloire and M. Vetterli, "Adaptive filtering in subbands with critical sampling: Analysis, experiments and application to acoustic echo cancellation," *IEEE Trans. Signal Processing*, vol. 40, pp. 1862–1875, Aug. 1992.

60. G.-O. Glentis, K. Berberidis, and S. Theodoridis, "Efficient least squares adaptive algorithms for FIR transversal filtering: a unified view," *IEEE Signal Processing Mag.*, vol. 16, pp. 13–41, July 1999.

61. G. H. Golub and C. F. Van Loan, *Matrix Computations*. Johns Hopkins University Press, Baltimore, MD, 1996.

62. E. Hänsler, "The hands-free telephone problem - an annotated bibliography," *Signal Processing*, vol. 27, pp. 259–271, June 1992.

63. E. Hänsler, "From algorithms to systems—it's a rocky road," in *Proc. IWAENC*, 1997, pp. K-1–4.

64. R. W. Harris, D. M. Chabries, and F. A. Bishop, "A variable step (VS) adaptive filter algorithm," *IEEE Trans. Acoust., Speech, Signal Processing*, vol. 34, pp. 309–316, Apr. 1986.

65. B. Hätty, "Block recursive least squares adaptive filters using multirate, systems for cancellation of acoustical echoes," in *Proc. IEEE ASSP Workshop on Application of Signal Processing to Audio and Acoustics*, 1989.

66. S. Haykin, *Adaptive Filter Theory*. Prentice-Hall, Englewood Cliffs, NJ, 1996.

67. J. Homer, I. Mareels, R. R. Bitmead, B. Wahlberg, and A. Gustafsson, "LMS estimation via structural detection," *IEEE Trans. Signal Processing*, vol. 46, pp. 2651–2663, Oct. 1998.

68. P. J. Huber, *Robust Statistics*. Wiley, New York, 1981.

69. International Telecommunication Union, "Digital network echo cancellers," Recommendation ITU-T G.168, 1997, section 3.4.

70. N. S. Jayant and P. Noll, *Digital coding of waveforms*. Prentice-Hall, Englewood Cliffs, NJ, 1984.

71. Y. Joncour and A. Sugiyama, "A stereo echo canceller with pre-processing for correct echo path identification," in *Proc. IEEE ICASSP*, 1998, pp. 3677–3680.

72. S. A. Kassam and H. V. Poor, "Robust techniques for signal processing: a survey," *Proc. IEEE*, vol. 73, pp. 433–481, Mar. 1985.

73. S. Kawamura and M. Hatori, "A tap selection algorithm for adaptive filters," in *Proc. IEEE ICASSP*, 1986, pp. 2979–2982.

74. S. M. Kay, "Some results in linear interpolation theory," *IEEE Trans. Acoust., Speech, Signal Processing*, vol. 31, pp. 746–749, June 1983.

75. S. M. Kay, *Fundamentals of Statistical Signal Processing: Estimation Theory*. Prentice Hall PTR, Upper Saddle River, NJ, 1993.

76. W. Kellermann, "Kompensation akustischer echos in frequenzteil-bändern," in *Proc. Aachener Kolloquium*, 1984, pp. 322–325.

77. W. Kellermann, "Kompensation akustischer echos in frequenzteil-bändern," *Frequenz*, vol. 39, pp. 209–215, Apr. 1985.

78. W. Kellermann, "Analysis and design of multirate systems for cancellation of acoustical echoes," in *Proc. IEEE ICASSP*, 1988, pp. 2570–2573.

79. W. Kellermann, "Some aspects of the frequency-subband approach to the cancellation of acoustical echoes," in *Proc. IEEE ASSP Workshop on Applications of Signal Processing to Audio and Acoustics*, 1989.

80. J. L. Kelly and B. F. Logan, "Self-adjust echo suppressor," U.S. Patent 3,500,000, Mar. 10, 1970 (filed Oct. 31, 1966).

81. L. M. van de Kerkhof and W. J. W. Kitzen, "Tracking of a time-varying acoustic impulse response by an adaptive filter," *IEEE Trans. Signal Processing*, vol. 40, pp. 1285–1294, June 1992.

82. S. G. Kratzer and D. R. Morgan, "The partial-rank algorithm for adaptive beamforming," in *Proc. SPIE Int. Soc. Optic. Eng.*, 1985, vol. 564, pp. 9–14.

83. S. M. Kuo and D. R. Morgan, *Active Noise Control Systems: Algorithms and DSP Implementations*. Wiley, New York, 1996.

84. J. C. Lee and C. K. Un, "Performance analysis of frequency-domain block LMS adaptive digital filters," *IEEE Trans. Circuits Syst.*, vol. 36, pp. 173–189, Feb. 1989.

85. L. Ljung, M. Morf, and D. D. Falconer, "Fast calculation of gain matrices for recursive estimation schemes," *Int. J. Control*, vol. 27, pp. 1–19, Jan. 1978.

86. V. Madisetti, D. Messerschmitt, and N. Nordström, "Dynamically reduced complexity implementation of echo cancelers," in *Proc. IEEE ICASSP*, 1986, pp. 1313–1316.

87. S. Makino, Y. Kaneda, and N. Koizumi, "Exponentially weighted step size NLMS adaptive filter based on the statistics of a room impulse response," *IEEE Trans. Speech Audio Processing*, vol. 1, pp. 101–108, Jan. 1993.

88. D. Mansour and A. H. Gray, Jr., "Unconstrained frequency-domain adaptive filter," *IEEE Trans. Acoust., Speech, Signal Processing*, vol. 30, pp. 726–734, Oct. 1982.

89. S. Marple, "Efficient least-squares FIR system identification," *IEEE Trans. Acoust., Speech, Signal Processing*, vol. 29, pp. 62–73, 1981.

90. R. Martin, "Spectral subtraction based on minimum statistics," in *Proc. EUSIPCO*, 1994, pp. 1182–1185.

91. R. Martin and S. Gustafsson, "The echo shaping approach to acoustic echo control," *Speech Commun.*, vol. 20, pp. 181–190, Dec. 1996.

92. J. E. Mazo, "On the independence theory of equalizer convergence," *Bell Syst. Tech. J.*, vol. 58, pp. 963–993, May/June 1979.

93. R. Merched, P. S. R. Diniz, and M. R. Petraglia, "A new delayless subband adaptive filter structure," *IEEE Trans. Signal Processing*, vol. 47, pp. 1580–1591, June 1999.

94. D. G. Messerschmitt, "Echo cancellation in speech and data transmission," *IEEE J. Selected Areas Commun.*, vol. 2, pp. 283–297, Mar. 1984.

95. O. M. M. Mitchell and D. A. Berkley, "A full-duplex echo suppressor using center clipping," *Bell Syst. Tech. J.*, vol. 50, pp. 1619–1630, May/June 1971.

96. M. Montazeri and P. Duhamel, "A set of algorithms linking NLMS and block RLS algorithms," *IEEE Trans. Signal Processing*, vol. 43, pp. 444–453, Feb. 1995.

97. B. C. J. Moore, *An Introduction to the Psychology of Hearing*. Academic Press, London, 1989, chap. 3.

98. D. R. Morgan, "Slow asymptotic convergence of LMS acoustic echo cancelers," *IEEE Trans. Speech Audio Processing*, vol. 3, pp. 126–136, Mar. 1995.

99. D. R. Morgan, J. L. Hall, and J. Benesty, "Investigation of several types of nonlinearities for use in stereo acoustic echo cancellation," *IEEE Trans. Speech Audio Processing*, submitted.

100. D. R. Morgan and S. G. Kratzer, "On a class of computationally efficient, rapidly converging, generalized NLMS algorithms," *IEEE Signal Processing Lett.*, vol. 3, pp. 245–247, Aug. 1996.

101. D. R. Morgan and J. Thi, "A delayless subband adaptive filter," *IEEE Trans. Signal Processing*, vol. 43, pp. 1819–1830, Aug. 1995.

102. E. Moulines, O. Ait Amrane, and Y. Grenier, "The generalized multidelay adaptive filter: structure and convergence analysis," *IEEE Trans. Signal Processing*, vol. 43, pp. 14–28, Jan. 1995.

103. B. H. Nitsch, "Real-time implementation of the exact block NLMS algorithm for acoustic echo control in hands-free telephone systems," in *Acoustic Signal Processing for Telecommunication*, S. L. Gay and J. Benesty, eds., Kluwer Academic Publishers, 2000, chap. 4, pp. 68–80.

104. K. Ochiai, T. Araseki, and T. Ogihara, "Echo canceler with two echo path models," *IEEE Trans. Commun.*, vol. 25, pp. 589–595, June 1977.

105. D. V. Ouellette, "Schur complements and statistics," *Linear Algebra Appls.*, vol. 36, pp. 187–295, Mar. 1981.

106. K. Ozeki and T. Umeda, "An adaptive filtering algorithm using an orthogonal projection to an affine subspace and its properties," *Electron Commun. Japan*, vol. 67-A, pp. 19–27, 1984.

107. B. Picinbono and J.-M. Kerilis, "Some properties of prediction and interpolation errors," *IEEE Trans. Acoust., Speech, Signal Processing*, vol. 36, pp. 525–531, Apr. 1988.

108. J. Prado and E. Moulines, "Frequency-domain adaptive filtering with applications to acoustic echo cancellation," *Ann. Télécommun.*, vol. 49, pp. 414–428, 1994.

109. L. R. Rabiner and B. H. Juang, *Fundamentals of Speech Recognition*. Prentice-Hall, Englewood Cliffs, NJ, 1993.

110. M. Rupp, "The behavior of LMS and NLMS algorithms in the presence of spherically invariant processes," *IEEE Trans. Signal Processing*, vol. 41, pp. 1149–1160, Mar. 1993.

111. L. L. Scharf, *Statistical Signal Processing*. Electrical and Computer Engineering: Digital Signal Processing, Addison Wesley, New York, 1991.

112. S. Shimauchi, Y. Haneda, S. Makino, and Y. Kaneda, "New configuration for a stereo echo canceller with nonlinear pre-processing," in *Proc. IEEE ICASSP*, 1998, pp. 3685–3688.

113. S. Shimauchi and S. Makino, "Stereo projection echo canceller with true echo path estimation," in *Proc. IEEE ICASSP*, 1995, pp. 3059–3062.

114. J. J. Shynk, "Frequency-domain and multirate adaptive filtering," *IEEE Signal Processing Mag.*, vol. 9, pp. 14–39, Jan. 1992.

115. D. T. M. Slock, "On the convergence behavior of the LMS and the normalized LMS algorithm," *IEEE Trans. Signal Processing*, vol. 41, pp. 2811–2825, Sept. 1993.

116. D. T. M. Slock and T. Kailath, "Numerically stable fast transversal filters for recursive least squares adaptive filtering," *IEEE Trans. Signal Processing*, vol. 39, pp. 92–114, Jan. 1991.

117. M. M. Sondhi, "An adaptive echo canceler," *Bell Syst. Tech. J.*, vol. 46, pp. 497–511, Mar. 1967.

118. M. M. Sondhi, "Echo canceller," U.S. Patent 3,499,999, Mar. 10, 1970 (filed Oct. 31, 1966).

119. M. M. Sondhi and D. A. Berkley, "Silencing echoes on the telephone network," *Proc. IEEE*, vol. 68, pp. 948–963, Aug. 1980.

120. M. M. Sondhi and W. Kellermann, "Adaptive echo cancellation for speech," in *Advances in Speech Signal Processing*, S. Furui and M. M. Sondhi, eds., Marcel Dekker, 1992, chap. 11.
121. M. M. Sondhi, D. R. Morgan, and J. L. Hall, "Stereophonic acoustic echo cancellation – an overview of the fundamental problem," *IEEE Signal Processing Lett.*, vol. 2, pp. 148–151, Aug. 1995.
122. M. M. Sondhi and A. J. Presti, "A self-adaptive echo canceller," *Bell Syst. Tech. J.*, vol. 45, pp. 1851–1854, Dec. 1966.
123. J.-S. Soo and K. K. Pang, "Multidelay block frequency domain adaptive filter," *IEEE Trans. Acoust., Speech, Signal Processing*, vol. 38, pp. 373–376, Feb. 1990.
124. A. Sugiyama, H. Sato, A. Hirano, and S. Ikeda, "A fast convergence algorithm for adaptive FIR filters under computational constraint for adaptive tap-position control," *IEEE Trans. Circuits Syst. II*, vol. 43, pp. 629–636, Sept. 1996.
125. M. Tanaka, Y. Kaneda, S. Makino, and J. Kojima, "Fast projection algorithm and its step size control," in *Proc. IEEE ICASSP*, 1995, pp. 945–948.
126. E. J. Thomas, "Some considerations on the application of Volterra representation of nonlinear networks to adaptive echo cancellers," *Bell Syst. Tech. J.*, vol. 50, pp. 2797–2805, Oct. 1971.
127. J. R. Treichler, C. R. Johnson, Jr., and M. G. Larimore, *Theory and Design of Adaptive Filters*. Wiley, New York, 1987.
128. V. Turbin, A. Gilloire, and P. Scalart, "Comparison of three post-filtering algorithms for residual acoustic echo reduction," in *Proc. IEEE ICASSP*, 1997, pp. 307–310.
129. P. P. Vaidyanathan, *Multirate Systems and Filter Banks*. Prentice Hall, Englewood Cliffs, NJ, 1992.
130. M. Vetterli, "A theory of multirate filter banks," *IEEE Trans. Acoust., Speech, Signal Processing*, vol. 35, pp. 356–372, 1987.
131. G. Wackersreuther, "On the design of filters for ideal QMF and polyphase filter banks," *AEÜ*, vol. 39, pp. 123–130, 1985.
132. A. Weiss and D. Mitra, "Digital adaptive filters: Conditions for convergence, rates of convergence effects of noise and errors arising from the implementation," *IEEE Trans. Information Theory*, vol. 25, pp. 637–652, Nov. 1979.
133. B. Widrow and M. E. Hoff, Jr., "Adaptive switching circuits," in *IRE Wescon Conc. Rec.*, 1960, part 4, pp. 96–104.
134. B. Widrow and S. D. Stearns, *Adaptive Signal Processing*. Prentice Hall Inc., Englewood Cliffs, NJ, 1985.
135. F. L. Wightman and D. J. Kistler, "Factors affecting the relative salience of sound localization cues," in *Binaural and Spatial Hearing in Real and Virtual Environments*, R. H. Gilkey and T. R. Anderson, eds., LEA Publishers, NJ, 1997, chap. 1.
136. M. Xu and Y. Grenier, "Time-frequency domain adaptive filter," in *Proc. IEEE ICASSP*, 1989, pp. 1154–1157.
137. S. Yamamoto and S. Kitayama, "An adaptive echo canceller with variable step gain method," *Trans. IECE Japan*, vol. E65, pp. 1–8, Jan. 1982.
138. H. Yasukawa, I. Furukawa, and Y. Ishiyama, "Characteristics of acoustic echo cancellers using sub-band sampling and decorrelation methods," *Electron. Lett.*, vol. 24, pp. 1039–1040, Aug. 1988.
139. H. Yasukawa and S. Shimada, "An acoustic echo canceller using subband sampling and decorrelation methods," *IEEE Trans. Signal Processing*, vol. 41, pp. 926–930, Feb. 1993.

140. H. Ye and B.-X. Wu, "A new double-talk detection algorithm based on the orthogonality theorem," *IEEE Trans. Commun.*, vol. 39, pp. 1542–1545, Nov. 1991.

Index

Printing: Saladruck, Berlin
Binding: H. Stürtz AG, Würzburg